Sylvia Weber Russell

Computer Interpretation of Metaphoric Phrases

Also of interest

Natural Language Processing and Cognitive Science: Proceedings 2014
Sharp, Delmonte (Eds), 2015
ISBN 978-1-5015-1042-7, e-ISBN 978-1-5015-0128-9, e-ISBN (EPUB) 978-1-5015-0131-9, Set-ISBN 978-1-5015-0129-6

Robots that Talk and Listen: Technology and Social Impact
Markowitz (Ed.), 2014
ISBN 978-1-61451-603-3, e-ISBN (PDF) 978-1-61451-440-4, e-ISBN (EPUB) 978-1-61451-915-7, Set-ISBN 978-1-61451-441-1

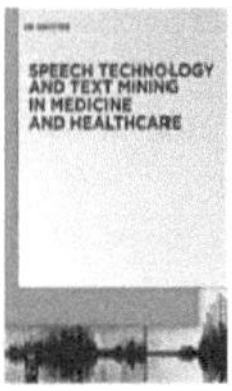

Speech Technology and Text Mining in Medicine and Healthcare
Amy Neustein (Series Ed.)
ISSN 2329-5198, e-ISSN 2329-5201

Sylvia Weber Russell

Computer Interpretation of Metaphoric Phrases

DE GRUYTER

Author
Dr. Sylvia Weber Russell
University of New Hampshire
Department of Computer Science
Durham
NH 03824 USA
swr@cs.unh.edu

ISBN 978-1-5015-1065-6
e-ISBN (PDF) 978-1-5015-0217-0
e-ISBN (EPUB) 978-1-5015-0219-4
Set-ISBN 978-1-5015-0218-7

Library of Congress Cataloging-in-Publication Data
A CIP catalog record for this book has been applied for at the Library of Congress.

Bibliographic information published by the Deutsche Nationalbibliothek
The Deutsche Nationalbibliothek lists this publication in the Deutsche Nationalbibliografie; detailed bibliographic data are available on the Internet at http://dnb.dnb.de.

Typesetting: Lumina Datamatics
Printing and binding: Hubert & Co. GmbH & Co. KG, Göttingen
Cover image: Gandee Vasan/Getty Images
© Printed on acid-free paper
Printed in Germany

www.degruyter.com

In memory of my parents, Alois and Margarete Weber
and for my husband, Bob

Preface

In the field of computational natural-language processing, metaphor has gone from being almost ignored to being a "hot-or at least warm-topic." My first foray into metaphor research was an outlier, one reason being that natural-language processing seemed difficult enough without attempts to interpret metaphor. In recent decades, there has been more awareness of the role of metaphor in what we consider ordinary or conventional language. Computational processing has followed suit, addressing mainly conventional metaphor, either with models which are partially semantics-based, or alternatively, with probabilistic approaches.

However, metaphor can also enable us to see things in a new or unconventional way. My intent in writing this book is mainly to suggest the extent to which a semantics-based model with a simple mechanism could (approximately) paraphrase both conventional and novel metaphor. But I hope that the book will also: provide an introduction for people outside of the field of natural-language processing who have an interest in factors underlying metaphor understanding; point to common ground and differences between my model and those of other researchers; and at least partially address the question of limits to computational analysis of metaphor.

My own metaphor research started through the back door. As I worked with Roger Schank's conceptual dependency representations for natural-language processing as a graduate student decades ago, it became obvious to me that these representations, while at an appropriate "conceptual" level for an interlingua for various purposes including paraphrase and translation, could be factored differently. That is, a slightly more "abstract" or reductionist view could show analogies that revealed themselves in conventional language-specifically through structural aspects that are invariant across physical and nonphysical domains. Such analogies were demonstrated early on by linguists as well, though with limited scope, as described in Chapter 3. With this answer in search of a question, I landed serendipitously on metaphor, and much of my work on verbal metaphor dates from this time.

I say "serendipitously," because analyzing metaphor was not just another obvious aspect of computational language processing. If the recognition and exploitation of patterns is a mark of creativity, then it seems that no cognitive phenomenon has wider creative application than metaphor and analogy, which is potentially involved in everything from scientific models through dreams to theatric plot invention. (Sadly, I have no qualifications to include these in this book.) Aside from a couple of other research detours, then, metaphor has been a continually interesting topic for me.

Most other computational approaches to metaphor in recent decades have focused on conventional metaphor, some of which efficiently use knowledge of known metaphors or of specific domains to interpret metaphors previously unseen, but with less attention to metaphor that can be considered novel. As the analyses in this book serve the goal of a wide scope, rather than being constrained to one or two domains, the material presented in the book is not a blueprint ready for general application. With its "breadth rather than depth" orientation with respect to domains, it is based rather on an interest in how metaphors work, whether embedded in everyday language or creative. The method, then, rather than relying on human or computerized 2 observations of usage, is one of systematic determination of elements of metaphor that can be formalized for basic literal paraphrases. The given paraphrases generated from the basic symbolic components may reveal what is missing from the interpretation, suggesting to any language user which components work and which do not.

With an interdisciplinary focus, the observations are often elementary and the style only minimally technical. Following general observations on metaphor and its processing by computers in Chapter 1, Chapter 2 shifts to a picture of the field of semantics-based computational methods of other researchers. Readers not familiar with computational metaphor analysis efforts may choose to leave Chapter 2 until the end or as reference when the work of these other researchers is used for comparison in later chapters.

Chapter 3 turns to my earlier research to include both the outline of the model to be described and some linguistic research that is relevant to it. In preparation for a narrower focus on cross-modal metaphors, which involve nonphysical domains, Chapter 4 presents the role of abstraction, with its interesting connections to cognition and mathematics. The representations and interpretation process for verbal phrases of this type are presented in Chapter 5. Representations for verbs described in Chapter 5 are used in Chapter 6 to represent predicates that are potentially salient to the interpretation of metaphorically used nominals.

Chapter 7 turns to consideration of a different type of figurative language, idioms, using representations similar to those of metaphoric phrases to interpret both simple and novel idiom modifications. Chapter 8, given the perspective of the presented work, offers a consideration of the extent to which metaphoric interpretations are computationally possible.

Both individuals and organizations have helped to support pieces and earlier versions of this work. I am indebted to Roger Schank, without whom none of this work would have happened. Robert Hoffman promoted my early interest in metaphor across disciplinary lines. The U.S.-Germany Fulbright program supported the rewarding collaboration I experienced at the University of Erlangen – Nuremberg, working on metaphoric idioms with Ingrid Fischer and Ricarda Dormeyer. I am grateful for the support of Prof. Hans Juergen Schneider of the Universty of Erlangen-Nuremberg, Dan Fass, and the Computer Science Department of the University of New Hampshire. Sons Patrick and Kevin deserve an enthusiastic mention. A nod also to the stimulating (pun intended) working environment of Breaking New Grounds in Durham. Finally, and most of all, thanks go to Bob Russell for his support in all possible ways.

Contents

1 Metaphors: Human Use and Computer Processing

In the early days of computational linguistics (natural language processing/NLP or natural language understanding/NLU), interpretation of metaphoric text was not a great concern. It seemed that there were enough difficulties to occupy researchers in the interpretation of "ordinary" language. Nevertheless, in the mid-1970s there was some implemented computer research on partly metaphoric sentences, [20, 27], and figurative language including metaphor began to be a "hot topic" in some areas of psychology and other disciplines. Ortony's edited multidisciplinary book, "Metaphor and Thought," [14] appeared in 1979; Honeck and Hoffman's edited book, "Cognition and Figurative Language" [9] and Lakoff and Johnson's book, "Metaphors We Live By," [12] followed in 1980. Lakoff and Johnson's book illustrated many "conceptual metaphors" – metaphor themes or formulas – that pervade our language, often or usually without our awareness. This explicit demonstration set many researchers in NLP to work on analyzing (mainly conventional) metaphoric language in text, frequently through reference to identified conceptual metaphors.

1.1 Views of metaphor

A typical dictionary definition of metaphor is that it is a figure of speech in which a concept in one domain is referred to as if it were another concept in a different domain. For our purposes, metaphor represents a "topic" in terms of properties extended from a "vehicle" in another domain. The topic is often referred to as the "target" and the vehicle as the "source;" less often, the topic is referred to as the "tenor." [19]. Here the terms "topic" and "vehicle" will be used, in order to avoid occasional potential confusion with other contexts of the words "source" and "target." The vehicle concept is represented by the metaphorically used word(s); the topic concept by a word or words signifying what is actually being described through the metaphor. In Shakespeare's "All the world's a stage," "stage" is the vehicle and "all the world" the topic. In the metaphoric "war horse," "The ship plowed through the sea/waves," "plowed" or "plowed through" is the vehicle and "ship" and "sea" belong to the topic domain. In its *metaphoric* sense, the verb in such an expression is also (made to be) seen as being in the topic domain.

Traditionally, three different views of the roles of topic and vehicle in metaphor have been recognized and debated, mainly in the fields of philosophy, psychology and rhetoric. These views have been variously understood, accepted and implemented. In the "substitution view," a word(s) used metaphorically is simply used in place of a literally used word(s) with the same meaning. For example, "He meandered into a buzzsaw of criticism" means the same as "He was heavily criticized." This idea would imply that these are equivalent, and that anything else supplied by the metaphor is "decoration" – a notion generally rejected today, e.g., by Verbrugge [26]. According to this view, however, a literal computer interpretation might serve as an equivalent.

In the "comparison view," an underlying similarity between topic and vehicle is assumed, which arguably applies to just a subset of metaphors, in particular those that exhibit physical similarity, such as "the highway snakes through..." or "the highway is a snake," in contrast to, e.g., "She buried herself in her work," or "Dreams are gold mines."

This view and the substitution view are not mutually exclusive, and the comparison view may be seen as a special case of the substitution view. Both views assume literal equivalence to the metaphor in what is being said. The comparison view, relying on similarity between topic and vehicle, potentially says more about the vehicle, here "snake." However, a similarity may be *created* rather than assumed, in which case the comparison view no longer seems appropriate.

Black [3, 4], influenced by Richards [19], promoted the "interaction" view, in which topic and vehicle interact, producing an image with characteristics common to both – the "common ground" of the metaphor. As the topic and notably, the vehicle are systems (an "implicative complex") rather than simple objects, the result of their interaction reveals something new, changing the way a receiver thinks about the topic (and in some cases the vehicle, if certain features are thereby reinforced). Since the originator of the metaphor may select (though not without constraints, if the metaphor is to work) various characteristics to apply to the topic, the interaction theory is not consistent with the comparison theory, is not a simple substitution, and is not subject to paraphrase. This view seems to accord with the attitude of Johnson (1982), who asserts that the existence of "preconceptual" or nonstructural elements of metaphor precludes modeling of the human ability to process metaphors.

However, there are metaphors which could be described by any one, or more than one, of these views. While the interaction view is sometimes characterized in different ways, it can be ventured that interpretation of the metaphor "Dumps are goldmines" works through the interaction view, as goldmines involves a complex of elements (yielding something valuable, "digging around," etc.) that could be productive in extending the "dumps" topic. On the other hand, while the comparison view might fail here, the interpretation simply that valuable things are found in dumps might support the substitution view. A metaphor that even better supports the interaction view would seem to be Shakespeare's "Juliet is the sun." While the effects of warmth and joy could be included in a paraphrase, the impact on Romeo of Juliet's/the sun's appearance is elusive. The interaction theory would seem to account best for all metaphors, though for an expression as conceptually simple as "the highway snakes..." it might be considered as "overkill."

These views relate more to other disciplines, and with the *notion* of metaphor than with computational processing; however, they do relate to researchers' own views of what their implementations represent. Most researchers appear to aim at achieving what the interaction view asserts to be involved in metaphor.

Metaphor is of general interest, however, not only because of its use in linguistic forms, but in other disciplines as well. Some aspects of linguistic metaphor correspond, e.g., to art. As Richards [19] asserts, a work of art is a medium of communication of an experience from the originator to the recipient. Linguistic metaphor may do this as well, and in turn may serve to make a poem, for example, have this experiential effect.

1.2 The metaphoricity of language

The fact that so much language considered literal has a metaphoric origin (even a simple phrase such as "I see what you mean" uses vision in a metaphoric way) motivates the question as to what is truly "literal" and what is metaphoric – a question of at least academic interest if we are interested in "literal" paraphrases of metaphor. This question will not be answered here, but has been extensively wrestled with in other disciplines, e.g., in psychology by Gibbs [8] and in the philosophy of language by Kittay [11]. Gibbs has asserted that distinctions between literal and metaphoric meanings "have little psychological validity." Ortony [15], by contrast, rejecting the "standard definition" of metaphor, which he states as "a word or phrase applied to an object or concept that it does not literally denote in order to suggest comparison with another object or concept" (p. 69), has claimed that a distinction can be made by reference to the context of the metaphor.

With respect to metaphor, literal meaning is sometimes designated as that unambiguous content which is extracted from the metaphor. In considering this notion, Gibbs, citing Allwood [2], points out that if literal meaning is that which is common to all contexts, then literal meaning is very general and abstract. This observation will be of relevance to our attempt at literal paraphrases of metaphor, since there must be something in common between metaphoric and literal usages.

It is sometimes stated that all language is metaphoric; language is a way of *re-presenting* the "real world" symbolically, in terms of categories such as "object" and "action." Bréal [5] stated that "language is a translation of reality" (p. 247, 1964 edition). In this sense, computer "primitive" symbols as well as words for spatial concepts could be considered metaphoric.

More restrictively, much language is etymologically metaphoric because we cannot directly represent nonphysical concepts. We resort to metaphorically based words to do so, such as the word "abstract" itself, derived from the concept meaning "draw away from." This and the preceding sense of "metaphor" are not relevant to the computational task.

In a further restriction, there are words such as "lose," which are used metaphorically so frequently that awareness of their metaphoric use is lost (the last phrase itself an example). Other examples are "give" and "see" (used for "understand"). Nominal concepts acted on by action concepts used metaphorically in this sense are represented as "objects," though they may not be physical. Some computational treatments of metaphor restrict themselves to this level of metaphor, or to usages which are recognizable as metaphoric, but are conventional, such as "kill."

Finally, there are expressions recognized by most readers or hearers as metaphoric, even novel.

Metaphor is therefore seen as a form of language creation, and the determination of "more literal" paraphrases as the stepping down from the "novel" or "unconventional" level to a "conventional" level, or from the "conventional" to a wordy literal paraphrase. This book will describe a metaphor interpretation program which translates metaphoric phrases into paraphrases which are "less metaphoric," and which the "person on the street" would consider literal. Both conventional and novel metaphor are treated.

1.3 Modeling metaphoric communication

The purposes of computational natural language processing include translation, text summary, information retrieval, data mining and other tasks. Metaphor of varying familiarity can occur in any one of these tasks, including speech processing [21].

"Natural language understanding" (NLU) by computer should itself not be understood in a literal sense. (As many computer-associated words of this type have become assimilated, however, no quote marks, unsurprisingly, will surround the occasional use of the word "understanding" in this discussion.) This notion of understanding may apply all the more to metaphor, as even human explanations of richer metaphors are usually unsatisfactory. The distant goal, then, is to produce computer paraphrases, not to reflect human understanding of metaphor, but rather to *approach* human paraphrases of metaphor.

While the output of a computer program is no proof of the theory or ideas that were implemented, it can make more specific the exploration of a mechanism or, often, indicate what does *not* work. Beyond the interest of computational linguists, however, the topic of metaphor has attracted a tremendous amount of interdisciplinary interest, including from the disciplines of psychology, cognitive linguistics, neuroscience, philosophy of language, literary work and criticism, social science and art.

1.3.1 What, how and why

It has been frequently recognized that metaphor can convey both information and emotion; other connotations may provide further nuances. How this is done is a question that is implicit in the work of many fields, including (to differing extents) some of the attempts presented in this book. In addition to factors proceeding from the views of metaphor described above, attention is given by various researchers to the role of affective factors and of semantic as well as pragmatic knowledge in the interpretation of the metaphor.

The "why" of metaphor brings us to the varied purposes of metaphor, which go far beyond the idea in earlier times that metaphor served only to add "decoration" to what was being stated. Decoration may play a minor role; there is some evidence [24] that in scientific articles more metaphoric words are used for general audiences than for audiences with more expertise, presumably to make the presentation more interesting. However, such metaphoric usage may also further comprehensibility.

Pollio, Barlow, Fine and Pollio [18] cite many intentions and purposes behind the use of metaphor. Purposes suggest that a comprehensive computational approach to interpreting metaphor will draw on existing NLU systems that represent goals and plans (a sequence of steps to achieve a goal) of speakers and hearers. In considering a metaphor, one might ask, for example, what the intent of the originator of the metaphor is, i.e., what the originator wishes to accomplish. Schank and Abelson [22] have designed systems that, given a text including a goal, can refer to plans to achieve that goal in order to fill in information missing from the text. Analogously in concept, a person with a goal can choose a metaphor that highlights what he or she wishes to convey. Conversely, a recipient of a metaphor might infer the goal from what is highlighted in a metaphor.

Perrault, Allen and Cohen [17], Perrault and Allen [16] and Allen and Perrault [1] base their analysis of indirect speech acts (considered by some to include metaphor, since metaphor is not literal) on the idea that the purpose of conversational participants is to have some effect on each other, in the form of an action or a new attitude.

The program presented from Chapter 3 onward does not include implementation that refers to plans and goals. However, the analysis does include the effect on the recipient of the metaphor and, implicitly, the intent of the originator. This approach is consistent with a view of meaning discussed more formally by MacKay [13], and serves the more general computational aim of analyzing metaphor in terms of human interests and goals.

Purposes of metaphor as a particular choice of language, then, go beyond brevity or just another way of saying something. Experiential components of a metaphor may reflect such purposes, including ulterior motives. Lakoff and Johnson [12] and others have observed that such components may not be able to be expressed through literal language.

1.3.2 Examples

A sense of how metaphor may call up an experience that may serve as a tool of intention can be given by evocative examples. For example, AFL-CIO's Lane Kirkland asserted that the Reagan Administration's budget was equivalent to "Jonestown economics," which "administers economic Kool-Aid to the poor." [10]. ("Jonestown" here refers to the place where a cult leader convinced large numbers of followers to drink poisoned Kool-Aid.) The negative emotions aroused by the remembered Jonestown incident is exploited by the speaker, who wants the listener/reader to perceive the administration's budget through the filter presented by the vehicle concept. His goal is to convince anyone listening (or reading about his assertions) to be against the Reagan budget. An adjective such as "cruel" to describe the budget would not have had this effect. The metaphor has made use of vehicle components, experience and negative affects to influence the listener. While the given metaphor might be thought by some to be in poor taste, Turbayne [25] viewed factors that potentially cause "shifts in attitude through highlighting or filtering" as the mark of a good metaphor.

Along the lines of persuasion, it should also be noted that, as Thibodeau and Boroditsky [23] have shown, framing the topic of a metaphor in different ways will probably elicit different responses to the topic. Whether crime is a "virus" or a "beast," for example, tends to predispose members of a community toward social remedies or increased law enforcement respectively.

An example of amusing imagery through metaphor (i.e., a simile, but the imagery created in this case is similar) appeals to humor [6]. In this scenario, a mother inside the house observes her children exhibiting rough behavior in the yard. Every few minutes, she bursts open the door "like a cuckoo clock" with the announcement that she has "just about had enough of you kids!" The scene could have conveyed about the same information in terms of the mother frequently coming out of the house to reprimand the children. However, not only does the cuckoo clock, which performs so regularly that one becomes accustomed to–and essentially unaware of–it, implicitly convey the impression that the children are oblivious to their mother's futile attempts, but the reader, perhaps along with the children, shares in the image which merges the mother with the cuckoo clock. This image in turn helps the reader to see, in a humorous way, how ineffective the mother's measures really are–in accordance with the goal of the writer. A literal human paraphrase could capture what happens, in a wordier way, but would not be as effective in terms of impressions. A computer program encounters at least as much challenge in its paraphrase attempts–and would be hard pressed to note the humor.

Metaphors and similes may also entertain in other ways, perhaps motivating the reader to dwell on an image. Consider a sentence by Garfield [7] (cited by Townsend [24]), describing a house with passageways among which were to be found single steps "that seemed like spies from lost battalions, lying in wait and wondering where the rest had gone." The reader's reaction to the merging of "wondering spies" and steps that are disorderly or simply misaligned is uncertain; he or she may absorb the image or be amused (or in this case may laugh *at* the simile).

It is generally recognized by metaphor researchers that a metaphoric vehicle can be used in the preceding ways to highlight a concept or experience that an originator wishes to share, and that certain perceived or imposed isomorphisms between topic and vehicle can be exploited to accomplish this. NLU researchers appear to have had some success with this aspect of metaphor, though sometimes giving only a few illustrations. The previous examples have further shown that appeals to the recipient's attitudes toward familiar incidents or experiences sometimes serve the intentions of the metaphor's originator. The "vagueness" that is often cited in emphasizing the difficulties in modeling metaphor derives partly from this aspect of metaphor, which should be incorporated when identifiable.

The expected result of describing a computer model of metaphor including all of the above elements is not complete interpretation, but rather an indication of how metaphor comprehension *might* be done–or not done. In these efforts, the main challenge is presented by vehicle salience determination and semantic representation–of both information and connotations–rather than by program mechanisms.

1.4 Outline

This hedging introduction to computational efforts to interpret metaphor notwithstanding, there is still much interesting progress to be made, whether for the purpose of information processing or for the sake of studying metaphor or linguistic communication itself. This book is an exploration through a semantics approach, claimed to be of more interdisciplinary interest than statistical methods.

The chapter which follows demonstrates a variety of approaches, but the differences are mainly in the type of metaphor and the methods used, such as differing levels of representation. Many of the assertions about metaphor and the principal mechanism involved, namely mappings, are shared in common. Most of the work presented is of course computational, but the analyses may themselves be of interest to those who study or use metaphor in other capacities.

Chapter 3 delineates various textual formats of metaphor, presents elements that can be used to describe verbs that would facilitate metaphoric interpretations, and introduces a computational approach to paraphrasing verbal metaphor through an early pilot program implementing these elements. In Chapter 4, a rationale is given for the creation of cross–modal or cross–domain metaphor, namely the formulation of a verbal or attributive concept as an object. To interpret cross–domain verbal metaphors, abstraction from verbs used with such objects is described in terms of an "abstract ontology." Chapter 5 explicates an abstraction-based program to paraphrase cross–domain verbal metaphor as a further development of earlier efforts. Included is a rough capability to distinguish metaphor from "incoherent" expressions. In Chapter 6, the more complex – and in some respects more interesting – task of paraphrasing nominal metaphor, together with metaphoric nominal compounds, is addressed.

As a large class of idioms is metaphoric, Chapter 7 presents a system to interpret modified idioms on the basis of meanings of their unmodified forms. Some thoughts on computational efforts to paraphrase metaphor constitute the conclusion.

References

[1] Allen, J., Perrault, C.: Analyzing intention in utterances. *Artificial Intelligence* **15**, 143–178 (1980)

[2] Allwood, J.: *On the distinctions between semantics and pragmatics*, pp. 177–189. D. Reidel Publishing, Boston, MA (1981)

[3] Black, M.: *Models and Metaphors*. Cornell University Press, Ithaca, NY (1962)

[4] Black, M.: More about metaphor, pp. 19–43. In: A. Ortony (ed.) *Metaphor and Thought*, Cambridge University Press, New York, NY (1979)

[5] Bréal, M.: *Semantics: Studies in the Science of Meaning*. Dover, New York, NY (1900). Later edition, 1964

[6] Dreikurs, T.: *Children: The Challenge*. Hawthorn, New York, NY (1964)

[7] Garfield, L.: *The Strange Affair of Adelaide Harris*. Longman Group Limited, current ed. Farrar, Straus and Giroux, New York (1971)

[8] Gibbs, R.: Literal meaning and psychological theory. *Cognitive Science* **8**, 275–304 (1984)

[9] Honeck, R., Hoffman, R. (eds.): *Cognition and Figurative Language*. Lawrence Erlbaum Associates, Hillsdale, NJ (1980)

[10] King, S.: Afl–cio's kirkland meets bush, attacks 'jonestown economics'. *New York Times* (1982). February 16, 1982, p. 1

[11] Kittay, E.: *Metaphor: Its Cognitive Force and Linguistic Structure*. Oxford University Press, New York, NY (1987)

[12] Lakoff, G., Johnson, M.: *Metaphors We Live by*. Chicago University Press, Chicago, IL (1980)

[13] MacKay, D.: *Information, Mechanism and Meaning*. MIT Press, Cambridge, MA (1969)

[14] Ortony, A. (ed.): *Metaphor and Thought*. Cambridge University Press, New York, NY (1979)

[15] Ortony, A.: Some psycholinguistic aspects of metaphor. In: R. Honeck and R. Hoffman (eds.) *Cognition and Figurative Language*, pp. 69–83. Lawrence Erlbaum Associates, Hillsdale, NJ (1980)

[16] Perrault, C., Allen, J.: A plan-based analysis of indirect speech acts. *American Journal of Computational Linguistics* **6**, 167–182 (1980)

[17] Perrault, C., Allen, J., Cohen, P.: Speech acts as a basis for understanding dialog coherence. In: *Proceedings of the Second Conference on Theoretical Issues in Natural Language Processing*, pp. 125–132 (1978)

[18] Pollio, H., Barlow, J., Fine, H., Pollio, M.: *Psychology and the Poetics of Growth*. Lawrence Erlbaum Associates, Hillsdale, NJ (1977)

[19] Richards, I.: *The Philosophy of Rhetoric*. Oxford University Press, London, England (1936)

[20] Russell, S.W.: Computer understanding of metaphorically used verbs. *American Journal of Computational Linguistics* **Microfiche 44** (1976)

[21] Russell, S.W.: Map: An abstraction-based metaphor analysis program for overcoming cross–modal challenges. In: A. Neustein, J. Markowitz (eds.) *Where Humans Meet Machines: Innovative Solutions to Knotty Natural Language Problems*. Springer Verlag, Heidelberg/New York (2013) pp. 163–183

[22] Schank, R., Abelson, R.: *Scripts, Plans, Goals and Understanding*. Lawrence Erlbaum, Hillsdale, NJ (1977)

[23] Thibodeau, P., Boroditsky, L.: *Metaphors We Think with: The Role of Metaphor In Reasoning*. url http://www-psych.stanford.edu/lera/papers/crime-metaphors.pdf (2011). Accessed 23-05-15

[24] Townsend, J.R.: *Written for Children*. Penguin Books, Ltd., Harmondsworth, Middlesex, England (1974). Citing L. Garfield, "The Strange Affair of Adelaide Harris," Longman Group Limited, England (1971)

[25] Turbayne, C.: *The Myth of Metaphor*. University of South Carolina Press, Columbia, SC (1971)

[26] Verbrugge, R.: Transformations in knowing: A realist view of metaphor. In: R. Honeck and R. Hoffman (eds.) *Cognition and Figurative Language*, Lawrence Erlbaum Associates, Hillsdale, NJ, pp. 87–125 (1980)

[27] Wilks, Y.: Making preferences more active. *Artificial Intelligence* **11**, 197–223 (1978)

2 Computational Models of Metaphor

Computational research on metaphor has taken various approaches, some of which relates to work by linguists, psychologists and cognitive scientists as well. Frozen metaphors can be retrieved directly from the lexicon; however, methods which depend on analysis to understand "live" metaphors bring along with them the capability of analysis of frozen metaphor as "part of the package."

From the point of view of focus, some work analyzes a particular domain in depth, while most research looks at metaphoric expressions in general, and are necessarily more open-ended. Metaphor themes which extend beyond a single sentence have not been given as much attention. Most of the work has been done on simple sentences in which the verb or predicate nominal, such as "warts" in "billboards are warts," [36] is used metaphorically.

The following sections divide research into verbal and nominal metaphor, though some researchers analyze both. While most of the work presented here is computational, some related work which has been influential is also included.

2.1 Verbal metaphor

Metaphor in which only a part of a sentence or phrase is used metaphorically is considered here in terms of whether the metaphorically used word(s) is a verb (or other predicative form), a noun or an adjective. Verbal metaphor is in turn partitioned into sentences in which both verbs and their object nominals[1] are physical, i.e., both in a physical domain, and those in which either the verb or its object nominal(s) is nonphysical.

2.1.1 Physical-domain metaphor

As nonstatistical computational research on metaphor is done in terms of some kind of analysis, it is reasonable to preface computational work by an indication of psychological research relevant to such analysis. **Gentner** [19] has conducted experiments on verb recall which indicates that verbs are stored as interrelated sets of components, suggesting that component properties and, especially,

1 The word "nominal" as used in this book is roughly equivalent to "noun," but does not include gerunds, such as "swimming."

relations rather than verbal concepts as a whole are extended in metaphor. In particular, it is relations that are principally salient in metaphors and analogies. This result is relevant not only to verbal physical–domain metaphor, but also to cross–domain metaphor, both verbal and nominal.

In Gentner's science–related physical–domain analogy, "The atom is analogous to the solar system," for example, the entities in the atomic and solar systems participate in corresponding relations: sun/nucleus ATTRACTS planet/electron, planet/electron ATTRACTS sun/nucleus, sun/nucleus MORE MASSIVE THAN planet/electron, planet/electron REVOLVES AROUND sun/nucleus. The simple attributes of the sun, YELLOW, HOT and MASSIVE, however are not extended to the nucleus of the atom. The transfer of structure posited for verbal metaphor plays a varying role for researchers of physical- and cross–domain metaphor.

Falkenhainer, Forbus and Gentner [13] have produced a computer implementation of the "structure-mapping engine" (SME) algorithm on the basis of Gentner's structure-mapping theory. In mapping relations in SME, they emphasize systematicity; that is, relations belonging to a systematic relational structure are preferred to isolated relationships. Systematicity is in contrast to literal similarity, in which both relational predicates and objects are mapped, and to "mere appearance," in which object descriptions are principally mapped.

SME uses descriptions of the topic and vehicle to construct all structurally consistent mappings between them. An interpretation is then a maximal structurally consistent collection of weighted matches according to certain rules. The "best" interpretation according to structural criteria is determined by a score of each interpretation on the basis of combined evidence for the individual match hypotheses.

Although the focus of Falkenhainer et al. is on analogy, aspects of SME are of obvious relevance to metaphor analysis. Gentner [20] emphasizes that explanatory analogies are similar to metaphors in that metaphors, being based on perceived analogies, can also be analyzed in terms of structure-mapping, although the similarity is less clear for "expressive"metaphor.

Along with the model to be described in later chapters, the computational research of **Wilks** [53, 54] on physical–domain verbal metaphor, followed by that of **Fass and Wilks** [15], was among the first to offer a procedure for interpreting verbal metaphor, using Wilks's "preference semantics." Wilks represents simple physical–domain sentences including metaphorically used verbs in terms of "primitive" case-based structures.[2] Every verb has a particular case configuration.

2 "Cases" refer to the roles (actor, object, etc.) that nouns or noun phrases used with a verb "fill" in its underlying structure. Case theory from a syntactic perspective was first proposed by

Thus he represents the verb "drink" as a *preferably* ANIMATE SUBJECT (agent) CAUSING to MOVE a liquid OBJECT TO a particular ANIMATE aperture (THRU PART) and INTO the SELF (the ANIMATE agent). He complements this information with information supplied by knowledge structures associated with definitions of contextual nominals. The partial knowledge structure for "car" includes the information that a MAN injects a #liquid into the car, USING a #tube, with the GOAL of making the MAN IN the car MOVE. The ICengine (internal combustion engine) USES the #liquid, CAUSING the car to MOVE, which CAUSES the MAN IN the car to MOVE. (Words in uppercase letters denote "primitive" concepts.)

For the metaphoric phrase, "My car drinks gasoline," "drink" in the input indirectly matches "inject" in the knowledge representation: both are CAUSE (to) MOVE verbs having the case-based inference that the object (liquid, gasoline) is IN the car. In order to come up with the literal meaning (beyond the fact that the gasoline is in the car), the procedure looks for the structure in the knowledge representation which is the closest match to the input structure. As ICengine is a part of a car and may substitute for it, and #liquid matches the object in the source structure ("gasoline"), the matching structure is "IC Engine (USE) #liquid." USE can then be projected onto "drink," producing "My car uses gasoline." Wilks's system, then, heuristically arrives at an interpretation by using world knowledge to change verb sense representations, thus assuring that the resulting interpretation makes sense.

In later research, Fass and Wilks [15] address further complications of metaphoric input of this kind. For example, the strategy used to interpret "My car drinks gasoline" did not succeed in interpreting the additional phrase in "The car drank gasoline and purred to itself." To resolve this problem, they employ two other strategies. For the first clause, "The car drank gasoline," a "change–the–expectation" strategy changes the expected subject of of the *verb*, i.e., "drink," from ANIMATE to VEHICLE. An alternative "change–the–data" strategy changes the subject *nominal* "car" from VEHICLE to ANIMATE. Because the unaltered subject *nominal* (VEHICLE) in the former representation leads to a "preference violation" (the preferred subject of "purred" is ANIMATE), the missing subject nominal for "purred to itself" comes from the representation which points to the latter strategy. This gives the representation "car (ANIMATE) –> (purred (SUBJ ANIMATE))."

A more comprehensive approach to this and other types of metaphor, as well as metonymy and other linguistic phenomena, is presented by Fass [14].

Fillmore [18] and reformulated from a "conceptual" perspective, e.g., by Schank [41] through his conceptual dependency theory.

Fass's "collative semantics," implemented by his *meta5* program, distinguishes sentences containing metaphoric uses of verbs from other linguistic phenomena in terms of one of seven types of semantic relation between pairs of words senses in the input. His interpretation of metaphor is in terms of comparison of the metaphoric statement with its literal counterpart based on the (literal) preference of the verb.

In meta5, the identification of a metaphoric relationship between a verb and its subject depends on 1) a preference violation: i.e., the subject nominal of the verb is not the expected (literal) one; neither is the expected subject its ancestor, as determined by a path through a word sense network; and 2) an analogical match between the relevant descriptor (called a "cell") of the expected subject (cell which contains the verb) and the actual subject; e.g., there is a "sister sense"–network path between the relevant cell of the preferred nominal and a cell of the actual subject nominal. In further processing, similarities and differences among the matches of *non*relevant feature–like cells of topic and vehicle nominals are counted; the more differences, the more "apt"[3] (roughly, "interesting") the metaphor.

Thus for "My car drinks gasoline," car1 is paired with the verb drink2, violating the preference of drink2 for "animal1;" the relevant cell "animal1 drink2 drink1" is a sister cell to and therefore analogical to "car1 use2 gasoline1," both having the common ancestor expend1; and the proportion of similarities to differences is moderate.

For the example, "The ship ploughed the waves," the noun ship1 is paired with the verb plough2, which prefers as a subject the noun plough1. No appropriate network path between ship1 and plough1 is found, so there is a preference violation. The relevant cell, [plough1, plough2, soil1], is found and results in an analogical match with [ship1, sail2, water2]. The program construes the input to mean that both a plough and a ship move through a medium. Tests then yield various degrees of matches of nonrelevant cells of plough1 and ship1, of which 4 (out of 9) are the same, including, for example, both having the feature cell "distinct bounds." Representation of the output semantic relations is in terms of semantic vectors–sets of ordered one-dimensional arrays populated by numbers representing cell matches and the result of network path searches.

Fass's system includes one of the more thought–out implementations of physical–domain metaphor. A question regarding meta5's arrival at its interpretation concerns the problems various researchers have with matching. For the

3 according to Tourangeau and Sternberg [46].

example, "The ship ploughed the waves," the analogy is found by matching [plough1, plough2, soil] with [ship1, sail2, water2," presumably because of their sharing a common ancestor in the hierarchy, which leads to both a "plough" and a "ship" "moving through a medium" as the common ground. It is difficult to know, however, how "medium" itself is defined; does it include the press or internet? Of course, Fass knew how he was defining something, and "medium" probably does not include those concepts, as they are not (purely) physical. But names for hierarchical network nodes are problematic; not everyone, perhaps few, would relate the verbs "drink" and "use" as sister nodes sharing "expend" as a superordinate, as Fass does to handle the example "My car drinks gasoline."

Another question regards the danger of mistaking a metaphor for anomaly. **Iverson and Helmreich** [30], in *metallel*, an extension of meta5, treat "The car drank coffee" as a metonymy that refers to the passengers in the car. However, "The rug drank [the] coffee" could not be resolved in this way, leaving the task to a test for metaphor. While there is the requisite preference violation, it does not seem that meta5/metallel could identify an analogical match between "rug" and "animal." This phrase could be comprehended as a metaphor describing what happens to spilled coffee, however, by reason of a match between an ingested (and therefore contained) liquid and a containment cell describing "rug," which in the absence of an identified analogy would presumably (and arguably wrongly) be considered *non*relevant. It is interesting, though, that the very feature which allows the metaphor (containment, *as applied to "rug"*) would be considered by Tourangeau and Sternberg [46] to make it less "apt."

A final question concerns the matching of nonrelevant cells that plays a role in calculating nonrelevant–cell differences. It is not clear how reliable the *number* of matched nonrelevant cells is. It seems that any conclusion would depend on the granularity and relative importance of the cells, though "importance" (of the cell) appears to have been addressed by extensions provided by Iverson and Helmreich.

Matching also plays a role in **Indurkya**'s early work on verbal metaphor [29]. His matching of structures between topic and vehicle is particularly appropriate for the many instances of physical–domain metaphor which exhibit literal similarity. He uses the example, "The sky is crying," to illustrate his focus on structural coherence of mappings, i.e., consistency of structural constraints in topic and vehicle domains. The tension between subject and verb is defined in terms of different domains, rather than literal preference violations. Here "cry" is in the EMOTIONAL–STATE domain and "sky" is in the WEATHER domain. Domain knowledge is represented as lists of statements in predicate calculus, as follows (the construct "X θ Y" indicates an exclusive "OR," i.e., "X or Y but not both"):[4]

DOMAIN 5: EMOTIONAL–STATE

1. cry(eyes) –> *fall–down* (tear–drops, eyes)
2. cry(eyes) <–> cry(person) AND *part–of* (person, eyes)
3. cry(person) –> very-happy(person) θ very-sad(person)
4. laugh(person) –> very-happy(person)
5. very–happy(person) –> NOT very-sad(person)
6. {*blue* (eyes) θ *dark* (eyes) θ *brown* (eyes)} AND NOT {*blue* (eyes) AND *dark* (eyes) AND *brown* (eyes)}
7. *liquid* (tear–drops)
8. *heavy* (tear–drops)

DOMAIN 6: WEATHER

1. *blue* (sky) <–> sunny(sky)
2. *blue* (sky) <–> day AND *clear* (sky)
3. sunny(sky) <–> day AND *clear* (sky)
4. *gray* (sky) <–> day AND cloudy (sky)
5. starry(sky) –> night AND *clear* (sky)
6. *contains* (sky, full–moon) –> night AND *clear* (sky)
7. *contains* (sky, full–moon) –> NOT starry(sky)
8. raining(sky) –> cloudy(sky)
9. raining(sky) –> {*gray* (sky) AND day} OR {*dark* (sky) AND night}
10. raining(sky) <–> *fall–down* (rain–drops, sky)
11. snowing(sky) –> cloudy(sky)
12. snowing(sky) –> {*gray* (sky) AND day} OR {*dark* (sky) AND night}
13. snowing(sky) <–> *fall–down* (snow–flakes, sky)
14. *heavy* (rain–drops)
15. *liquid* (rain–drops)

---

4 Domains are from "A Computational Theory of Metaphor Comprehension and Analogical Reasoning" (Tech. Rep. No. 85/001, pp. 162–163) by B. Indurkhya, 1985, Amherst, MA: University of Massachusetts. Copyright 1985 by B. Indurkhya. Reprinted by permission.

16. *light* (snow–flakes)
17. *solid* (snow–flakes)

Indurkhya also gives some attention to "weakly coherent" metaphors, in which there is no evident literal similarity. In this area, his assignment of unmatched information from the vehicle domain to the topic is similar to the procedure discussed later in this book. The method he specifies in his 1985 work as follows, however, is oriented toward physical–domain metaphors exhibiting similarity.

The procedure begins by looking at constraints of EMOTIONAL–STATE containing "cry". Constraint 1 of EMOTIONAL–STATE, "cry(eyes) –> *fall–down* (tear–drops, eyes)" matches both Constraints 10 and 13 of WEATHER. Constraint 13 suggests the mapping "cry –> snowing); (eyes –> sky); (tear–drops –> snow–flakes)." Verification of the coherence of this mapping, however, reveals that Constraints 7 ("*liquid*(tear–drops)") and 8 ("*heavy*(tear–drops)") are contradicted by Constraints 17 and 16 of WEATHER, which specify that snow is "solid" and "light," respectively, so Constraints 7 and 8 are not transferable, and Constraint 13 is not applicable to this example.

If the context requires another interpretation, Constraint 10 of WEATHER is considered. In this case Constraint 1, 7, and 8 of EMOTIONAL–STATE *are* transferable, resulting in the interpretation "raining (sky)." Transfer of human attributes such as "sadness" could be facilitated by the existence of a token in WEATHER corresponding to "person" of EMOTIONAL–STATE. Indurkhya thus uses domain information, described to some detail, to discover similarities that might serve as the ground of a physical–domain metaphor. He does not explicitly posit primitive components as a factor in his matches, but some of the simple "constraints" he specifies to describe his domains, such as "fall–down," "liquid" and "heavy" can be seen as filling that role.

Indurkhya's domain descriptions contain all the information that can serve as the basis for interpretation of the given examples. There is some question as to the semantic content of his predicate calculus notation. For instance, the specification of "raining" as an attribute of "sky" seems forced; is "sky" an actor of raining, or is it the clouds, or is rain in itself an event, with the sky being the direction from which it comes? Not using more specific semantic relationships (e.g., as Fass and Wilks use) may be a representational disadvantage. This is a small example, but the information about "sky" comes from the WEATHER domain, which in an operational system would be much larger. In fairness, this metaphoric simple sentence is semantically problematic, as researchers have argued about how to represent "rain," and since "sky," rather than an objective concept, can be thought of as an entity conceived by an earth viewer for purposes of location/orientation.

The work of **Weber** [51] is almost unique in metaphor research in focusing on adjective–noun constructs. As these could alternatively have a predicative form, her work is included in this (verbal metaphor) section. Weber's (connectionist) system learns literal meanings first and refers to known meanings in its understanding of usages new to the system. Feature values of the adjectival concept are applied to salient features of the nominal concept.

For example, the sense of "green" as a color is learned first, followed perhaps by its linkage to "unripe," because of the tendency of unripe fruit to be green. With enough reinforcement, this sense may itself become established. With the "green–unripe" correlation and an abstraction linking an unripe quality with inexperience, the combination "green recruit" can be interpreted.

The structure of connections in the spreading activation network in Weber's system reflects "direct inferencing," which is the "automatic change in property value settings" on the basis of *one* property value. For example, that a "green banana" is "unripe, difficult to peel and astringent tasting" is stored with the "banana" concept, along with the default values of a yellow banana.

"Functional" properties provide the organization structure of each category. That is, the functional property "ripeness" gives rise to three "aspects" of bananas: unripe, ripe and overripe. In encountering new usages of adjectives in combination with a noun, the vehicle domain consists of the connotations of the adjective as used in previous usages with other nouns. The considered connotations are the immediate inferences triggered by changes to a category (noun) "commonly associated" with the adjective in question. When an adjective is applied to a category with which it is not normally associated, it is assumed that it is referring elliptically to a property of the noun with which it *can* be associated. For example, "thirsty" in "thirsty fern" refers to the fern's need for water, though the fern is not truly conscious of it.

Along with direct inferencing, scalar value transference anchors Weber's model. The model includes the processing of adjectives representing features which are scalar indications relative to a norm, such as "large" or "light," as well as features that are *collections* of scalar feature values, such as taste (sour, sweet, etc.). The former values are mutually exclusive, while the latter values can co–occur. Confidence values are attached to feature units.

The basis for unfamiliar figurative usage interpretation consists of the set of all values resulting from mappings from the vehicle domain (as defined above) to the values of the nominal concept. Semantic correspondences are established either by property value transference, through direct inferencing, when there is a property value shared by topic and vehicle, and an associated value can be directly inferred from the vehicle and thus transferred, or by scalar correspondence, e.g., "small" maps to "lightweight," which is handled by the property abstraction hierarchy.

It appears that some details of Weber's model still need to be worked out. For instance, she mentions that, "all else being equal," the salience of nominal properties would be favored in interpreting novel phrases such as "happy box," but does not go into what determines salience. Also, one would like to know whether processing of other types of adjectives in such phrases has been implemented, and what differences there might be. More research on this metaphoric form would be welcome.

2.1.2 Cross-domain metaphor

The system of **Hobbs** [23, 24] utilizes axioms formulated in terms of predicate calculus to represent metaphoric extension. "Schemata" comprising collections of axioms interrelated by the co-occurrence of some of the same predicates are used to represent complex novel metaphors. To illustrate, he uses an example from a Democratic congressman who complains about President Ford's vetoes [40]: "We insist on serving up these veto pitches that come over the plate the size of a pumpkin." The metaphor is represented as two linked schemata. In the schema representation "send (SD,C,B,P)," SD is the action by Congress C of sending the bill B to the President P; in the representation "pitch (*p*,*x*,*y*,*z*)," *p* is the pitching *p* by the pitcher *x* of the ball *y* to the batter z.

To determine the analogy, Hobbs attempts to match predicates attached to the metaphoric concepts with those derived from world knowledge about the topic (cf. Wilks's projection from a knowledge structure). Thus knowledge of baseball ("sending a ball to a pitcher") is extended to knowledge about vetoes ("sending a bill to the President").

In later work [25, 26], Hobbs revises the framework of his method by presenting interpretation as "abduction," or, roughly, reasoning back from the metaphor to explain it. That is, given a complex topic expression, this expression is decomposed into basic concepts in the vehicle domain; the analogy is then undone to arrive at the meaning in the topic domain.

He again illustrates with the above metaphoric sentence, in which the two schemata and the relation between sending and pitching are represented as axioms. Hobbs summarizes [25]:

> Assuming the Congress schema, assuming what was needed to infer a pitching from a sending, and assuming the missing pieces of the baseball schema, we derived the baseball schema and identified the corresponding actions, thereby interpreting the novel metaphor. The two schemas together then accounted for the propositional content of the sentence (p. 23).

Hobbs does not appear to have implemented his method, focusing rather on his proof.

Lakoff and Johnson [32], from a cognitive–linguistic perspective, have created an extensive list of what they call *conceptual metaphors*, such as *LIFE IS A JOURNEY* ("We have a rocky road ahead of us"), *THE MIND IS A CONTAINER* ("Keep it in mind," "Be open-minded"), *TIME IS MONEY* ("Don't spend your time on this") and *IDEAS ARE FOOD* ("I'll have to chew on this before I can swallow it"). These metaphors structure much of our language as frozen or dead metaphors; we are usually not aware of such metaphors when we use them without reflection. The pair of metaphors *LIFE IS A JOURNEY* and *LIFE IS A GAME* ("As a child, he was not dealt a good hand") is an illustration of how different vehicle concepts may be used to structure the topic in different ways.

Lakoff and Johnson's conceptual metaphors represent metaphor themes after they have entered into common usage; thus their metaphors are conventional rather than novel, but most of the conceptual metaphors appear to be cross–modal, created to convey concepts not easily expressed in the physical domain. Lakoff and Johnson did not organize their conceptual metaphors and they are not interpreted computationally. However, conceptual metaphors have served as a basis for the computational models of others.

Grady, Oakley and Coulson [21], in connection with the research of Fauconnier and Turner [16, 17], have from an interdisciplinary perspective both distinguished their "conceptual blending" framework from that of conceptual metaphors and suggested the complementarity of the two theories. While conceptual metaphors have become integrated into our language as frozen, they point out that blending allows "short–term" enrichments to conceptual metaphors in an "on-line" process through additional mappings. They use the term "mental space" in place of "domains" in referring to scenarios that are structured by given domains. Given a metaphoric expression with the topic in the intellectual domain and the vehicle in the visual domain (according to an entrenched conceptual metaphor, e.g., *UNDERSTANDING IS SEEING*), then, any/all relatively novel metaphors that are instances of, derived from or elaborative of the expression might be said to be in a mental space.

Huang, Huang, Liao and Xu [28] present a revised conceptual blending framework, which includes a lexical ontology, with context as a "fifth mental space." Metaphor mappings are based on the ontology mappings. In addition to the two domains of conceptual metaphors, they posit a "generic" space representing conceptual structure shared by both domains, and a "blend" space, where material from the input interacts and combines. They use the metaphoric expression, "The committee has kept me in the dark about this matter" to characterize the space for the visual domain as a person standing in the darkness and the other as a committee withholding information from the same person, and a mapping between these spaces. The generic space contains the shared material, approximately, “a person who has no access to a particular stimulus”; in the blended space, a committee causes an individual to remain in the dark.

Grady et al. also discuss a nominal metaphor, “This surgeon is a butcher,” for which there are fairly obvious mappings between the two, which, however, do not reveal that the surgeon is incompetent. Conceptual blending produces this meaning by including not only the differing goals of the surgeon and butcher, namely “healing” and “severing flesh” respectively, but also the conclusion that the destructive means a butcher uses (“butchery”) is inconsistent with the surgeon’s ends; therefore the surgeon is incompetent. The possibility of drawing on various scenarios derived from this metaphor is reminiscent of the “scripts” (and variations thereof) of Schank and Abelson [42]. With the acknowledgment of the complex relationships of conceptual metaphors that may occur in longer pieces of text, there is potential to analyze pieces of discourse as well.

Like Hobbs, **Carbonell** [6, 8] treats small pieces of discourse containing verbal metaphor but also involving metaphoric nominals. He states that a purpose of metaphor is to structure inferences, including goals and plans of the actors in the topic domain. Lakoff and Johnson’s stored conventional conceptual metaphors serve as a means of interpreting metaphorically used words and phrases.

For example, Carbonell uses the *MORE IS UP* conceptual metaphor to interpret the New York Times headline, “Speculators brace for a crash in the soaring gold market.” In Carbonell’s sytem, “soaring” (“upward movement”) is matched with “up” and “increase in value or volume” of “gold market” is matched with “more” in the *MORE IS UP* metaphor. Propositions attached to “soaring” can be transferred to “gold market:” “The increase is rapid and not firmly supported; if it tumbles it may undergo a negative state change.” Here metaphoric language is mapped from the physical vehicle domain to a financial topic domain. This mapping is applied to the analysis of the rest (first part) of the sentence: “Speculators brace [prepare] for a [physical/financial negative state change caused by a physical/financial] crash.”

These inferences are clearly analogical and the paraphrase sophisticated. It is not indicated how the actual choice of words derives from representations of the conceptual metaphor, or whether salience recognition plays a role in this part of Carbonell’s approach. Awareness that a stock market (which is in the topic domain) could “tumble” is a relatively strong possibility and therefore potentially salient. But “tumbling” in the physical domain is arguably of lower or at least not higher salience. This observation appears to contrast with Ortony’s [36] salience theory, where predicates in the vehicle are of high salience and in the topic of low salience. Yet the above example works, apparently because inferences or properties other than the one most salient to the metaphor as a whole are not subject to this rule.

Carbonell thus shows that metaphors can structure whole paragraphs. Furthermore, different goals follow from different metaphors, e.g., “defeating” or “curing inflation” from the *INFLATION IS WAR* or *INFLATION IS A DISEASE* metaphor, respectively. The process of determining which inferences should be transferred is supported by reference to salient properties ranked heuristically in an invariance hierarchy. [7, 8] The hierarchy was established by means of a study of about 200 examples.

In decreasing order of expected invariance, a sampling of the properties is as follows (derived from [8], pp. 19–20):

A goal–expectation setting for the animate actors involved (if any), as in the *INFLATION IS WAR* example.

Causal structures.

Functional attributes.

Temporal orderings.

Natural tendencies (water tends to go downhill; therefore electricity tends to go towards the voltage “drop”).

Structural relations (as observed by Gentner).

Object identity (almost never mapped).

In later work, **Carbonell and Minton** [9] use a transfer of parts of a graph consisting of concepts linked by relations to specify their analogy-based mechanisms. Thus for “X is a puppet of Y,” the concept node CONTROL between the object “puppet” and the actor “puppeteer” is transferred to the node between “X” and “Y.” Because there is no established conventional conceptual metaphor for the vehicle domain, the vehicle and topic domains are compared to create new mappings, a process constrained by the invariance hierarchy. This combined mechanism is used to interpret the example, “Bulgaria is a Russian puppet,” as “Russia controls Bulgaria,” no doubt the desired interpretation. In general, as the choice of predicate is determined mainly by the heuristic invariance hierarchy, we could say that this method, though with the possibility of missing the mark, has “a good chance” of achieving a reasonable interpretation.

Dyer and Zernik [11, 12], in their research on parsing, representing and acquiring figurative phrases, refer to an implemented dialog system, RINA, which as a subtask demonstrates some processing of metaphor. The phrases are idiomatic, usually with a metaphoric basis. By “acquiring,” they mean that if such a phrase is not in the phrasal lexicon, world knowledge and the immediate context of the phrase is used to produce an interpretive representation, and its meaning can be included in the lexicon.

As an example, they use "The Democrats in the house *carried the water for* Reagan's tax–reform bill."[5] RINA asks for confirmation of a literal interpretation, which is rejected. It then asks, "They helped him pass the bill?", which is confirmed. This hypothesis (and the meaning subsequently generated) is derived from salient features of the immediate context, in terms of scripts, relationships, plan/goal situations and emotions. In this case, (to simplify) there is reference to a means of *executing a plan* to achieve a *goal*, i.e. passing the bill. The goal is generalized for insertion of the phrase meaning into the lexicon, as something like "helping another person do a hard job," since not all contexts involve passing a bill.

It is relevant for metaphor interpretation that emotions and attitudes play a role in Dyer and Zernik's interpretations. Their treatment, however, differs from a direct attachment of an emotion to a metaphoric phrase; rather it is derived from plan/goal situations, such as the failure to achieve a goal. Elsewhere [10], in the context of ordinary narratives, Dyer defines several components used to trace emotion states of characters in a narrative, including scale, whether the emotion can be directed toward another, and expectations of characters about future results of a goal.

Martin [34] has implemented a system called MIDAS, which relies on pre-existing conventional metaphoric knowledge of the type collected by Lakoff and Johnson. This knowledge, represented as knowledge structures in the KODIAK language, resides in a lexicon together with word senses. In MIDAS, conventional metaphors are represented as coherent sets of links between vehicle and topic concepts. For example, the interpretation of the conventional (frozen) metaphor "Mary gave John a cold" utilizes the associations between components, i.e., between the giver and the infector and between the given object and the infection to produce the paraphrase, "Mary infected John with a cold."

Unknown metaphors are interpreted (in an expansion of MIDAS called MES) by an extension of an existing metaphor and added to the knowledge base as a form of "learning." Metaphors are extended through recognition of 1) *core relationships*, 2) *similarity* and 3) a *combination* of these two. Core relationships are those between a state, its beginning and ending, and other semantically close components, similar to case roles. Such relationships exist, for example, between "give," "have" and "get," a "giver" and a "givee." They can be thought of as derivative from the core metaphor, "Infection–As–Possession," of which "have" is a possible lexicalization. Similarly, "enter," "in" and "out" are core–related. Such relationships occur in metaphoric mappings between vehicle and topic, e.g., a "giver (of a cold)" is mapped to "infector;" the relationships are all

5 From the ABC TV program *Nightline*, 12/12/85

transferable from the vehicle to the topic. The mappings together comprise a new *metaphor sense*, such as "*Give*–A–Cold."

Similarity between metaphors is recognized through use of abstraction hierarchies of nominal concepts. The abstraction hierarchy includes *core metaphors*, which are general metaphors high in the hierarchy, and specialized metaphors, in which either the topic or the vehicle is subordinate to a core metaphor. If an unfamiliar metaphor is encountered, the system compares it to a similar known metaphor and tries to find the closest common abstract ancestor in the hierarchy. "Giving a cold" as a known metaphor, for example, can be referred to in interpreting "giving the flu" in terms of "infecting," since both "cold" and "flu" are dominated by the "Communicable Disease" concept in the abstraction hierarchy. As another example, the interpretation of the unfamiliar metaphor "Kill–Computer–Process" involves reference to a more distant closest ancestor–the category "Process," which has the subordinate categories "Computer–Process" and "Conversation." Because of this relationship, the known metaphor "Kill–Conversation" provides a basis for accepting the new metaphor "Kill–Computer–Process."

Martin also gives an example, "Using–Lisp," for using a combination of both core relationships and similarity to extend a metaphor. If the system knows the "In–Emacs" metaphor ("in" is used metaphorically here) as literally "Using–Emacs," it can understand "In–Lisp" by similarity extension: "In–Emacs" is understood as "Emacs–Active," which is abstracted to its ancestor, "Comp–Process–Active;" this ancestor can accept "Lisp" in the role played by "Emacs," since they are both subordinate to "Computer–Process," giving "Lisp–Active." Given the core relationship between "Lisp–Active" and "active Lisp *user*, then, the system could presumably understand being "In–Lisp" as "Using–Lisp."

Potential metaphoric interpretations are obtained through "metaphoric unviewing" procedures that retrieve topics from representations of conventional metaphors, such as "infecting" from *INFECTING AS GIVING*. The choice of interpretation depends on satisfaction of the constraints which the verb imposes on its associated nominals. The most specific topic (verb) satisfying those constraints is chosen.

The abstraction hierarchy used by MIDAS/MES appears to be word-oriented. In the absence of semantic analysis of the verb, it is unclear whether an interpretation based on the hierarchy would be found for, e.g., "snuff out" in place of "kill." Interestingly, MIDAS/MES does find a connection between "Inflation is eating up our savings" and "My car drinks gasoline." It would seem, though, that this analogical connection could also be made simply from the immediate inference on either "eat" or "drink" and their shared inference of loss of amount, not needing recourse to a necessarily large abstraction hierarchy or knowledge of a previous metaphor.

In any case, Martin's system appears successful at finding the analogies underlying many verbal metaphors. A question is whether the abstraction hierarchy would eventually become unwieldy. For example, is playing the piano (in Martin's own categories), a "Mechanical Process," like typing, or a "Human Activity"? If the latter, is it a form of "Game Playing" or a "Physical Activity," or perhaps a "Speaking Activity," since "Performance" is given as a type of "Speaking Activity"? Or do we need a new category of "Expression"? This is a question not particular to Martin's system; category hierarchies are used by other researchers as well.

Hayes and Bayer [22] focus on metaphors in which spatial concepts are extended to temporal concepts. They consider verbal forms, such as "The current calendar year overlaps with the next fiscal year," but also attributive forms, such as "The length of the meeting is five hours." The various space-to-time mappings, implemented as part of a lexicon, can also be linked together to interpret a sentence. For example, the literally spatial "length (of)" is mapped to a measure of time, and this is chained together with mappings of other constituents of the sentence. With this particular domain target, Hayes and Bayer can interpret many conventional space-to-time usages.

Lytinen, Burridge and Kirtner [33] have implemented a system, LINK, distinguished for its handling of metaphor, metonymy and idioms in one sentence. Faced with mixed psychological studies, they argue for the hypothesis that humans compute literal meaning of a metaphoric expression in their interpretation of metaphor, at least in some cases.

In its interpretation process, LINK constructs literal interpretations of the constituents of the expression, trying to match them with mapping rules. If the rules match, the result of the mapping is accepted as a possible nonliteral interpretation of the constituent. Finally, contextual information is used to select (an) acceptable interpretation(s).

Mapping rules are simple because of their productivity. The example, "The temperature rose," along with many other metaphors, is interpreted through a mapping rule which says (in simplified form) that CHANGE-ALTITUDE is mapped to CHANGE-VALUE. A metaphoric interpretation, then, is generated automatically in the case of a match with a mapping. The literal interpretation of the metaphorically used word appears to be used only in the sense of a mapping match rather than by semantic computations or violations. This process is consistent with the Standard Model (which involves mapping rules), which says that literal meaning is computed, but also with psychological studies that say that metaphoric comprehension is just as fast as literal comprehension. With the ability to parse and interpret several kinds of nonliteral (though not novel) language, LINK appears to be a small but forward-looking NLP program in itself.

Suwa and Motoda [44] use a "primitive–matching" constraint to select a topic–domain verb in interpreting cross–modal metaphor. Their approach accords with Ortony's [36] theory, in that non–salient features of a topic are highlighted through salient features of a vehicle. Thus they import structures from both topic and vehicle.

Verbs are defined in terms of "primitives" – "lower–order" (e.g., an action or state) and "higher–order" (e.g., causation).[6] Suwa and Motoda define the topic of a verbal metaphor as the verb that would be literal for the expression. As an example, for "He shot down my opinion," the verb "shoot down" in its literal sense is the vehicle. Its definition is matched against verb candidates that would serve as a topic, i.e., a verb defined with the maximum number of similar primitives, with "opinion" as a possible object. In this case the verb "criticize" is found. Suwa and Motoda's abstract representations and, to some extent, their method, appear to be similar to the model for verbal metaphor [40] described in Chapter 5.

The work of **Barnden et al.** [3, 4] and **Agerri et. al.** [2] recognizes that many metaphoric usages do not adequately correspond with Lakoff's conceptual metaphors. In order to extend components that are not part of any particular "metaphorical view" or conceptual metaphor [2], they posit *view-neutral mapping adjuncts (VNMAs).*

Barnden et al.'s views on mapping appear to be similar to the "conceptual domain crossings" [39] described in Chapter 3, including the extension of emotions, and many of the view–neutral mappings in the two systems agree. However, the ontology of Barnden et al. appears more comprehensive in that, for example, they include *Duration* and *Set–hood.*

To illustrate the use of VNMAs, Barnden et al. and Agerri et al. both use the example, "In the far reaches of her mind, Anne knew that Kyle was having an affair..." [1] They refer to the Lakovian *MIND IS PHYSICAL SPACE* and *IDEAS ARE PHYSICAL OBJECTS* conceptual metaphors, with their mappings, including, e.g., that if a person's mind is considered as a physical space, then an idea in the person's mind allows the person to consciously operate on the idea. The interpretation is implemented using the ATT–Meta "metaphorical reasoning" system, [5], which is rule-based. To (greatly) simplify, they presume the query:

to–degree–exactly(D):can–consciously–mentally–operate–on(anne,
the–idea–that(having–affair(kyle))).

---

6 These descriptions appear to resemble the causally linked, primitive–based "conceptualizations" of Schank's [41] Conceptual Dependency theory, though they may be more "abstract."

where D is the queried variable. This query is transferred from the mental to the physical domain; it is here, i.e., in the *vehicle* domain, where reasoning takes place:

IF to-degree-at-least(D): can-physically-operate-on(P,J)
THEN to-degree-at-least(D): can-consciously-mentally-operate-on(P,J)

The mechanism employed here appears somewhat opaque to operative analogies, but does capture mappings that are significant to metaphor interpretation.

The verbal metaphor theory underlying the system of **Narayanan** [35] is similar in some respects to the research of Barnden et al. and Agerri et al. Like other researchers, he posits spatio-temporal structures as extensible to other domains, but goes further to establish correspondences between motion verbs as part of a *sequence* of actions and inferences leading to a goal in a topic domain. Some elements of Narayanan's treatment of nominals, verbs and adverbs in terms of invariant components correspond with those [40] described later in this book. For example, from looking at various databases, Narayanan has confirmed the invariance of certain "parameters" which resemble adverbial features (refer section 5.2.2) expressing, e.g., evaluation and agent attitude/intent.

However, Narayanan focuses on just two topic domains, economic policy and politics, implementing his specific mappings (in a neural-like system) with fine granularity. Entities (nominals) and actions are projected directly onto the topic through stored conceptual metaphors in the sense of Lakoff and Johnson. With his (relatively!) narrow focus, Narayanan is able to achieve an effective system of verbal metaphor interpretation.

Finally, the work of **Shutova** [43] bears a mention, though her model is a probabilistic one, because her goal, namely literal paraphrase, which deviates from the output interpretations of most researchers, is similar to one proposed in this book. Shutova reports an accuracy rate of 81%.

2.2 Nominal metaphor

Nominal metaphor as described here is considered within-domain (physical-domain) not just because both nominals are physical, but because the predicates that enter into their interpretation are also physical, with the exception of the first example.

2.2.1 Physical-domain metaphor

Winston [55] introduces a theoretical rationale for determining salience for nominals. He states some principles that aid the selection of potentially salient properties from a definition of all the properties of a metaphorically used concept, namely, *importance* of a property, *extreme value* and *distinction with respect to the class* of the object.

These principles function as tentative "filters" in the form of "transfer frames" for properties to be transferred from the vehicle to the topic concept. Frames consist of a frame name, slot or attribute, and value. A vehicle frame for "fox," for example, named FOX, referred to for the metaphor "Robbie is a fox," has a slot A–KIND–OF with value SMALL–MAMMAL, a slot COLOR with value RED, and a slot CLEVERNESS with value VERY–HIGH. CLEVERNESS: VERY-HIGH in the transfer frame then serves as an extreme value.

The vehicle frame for a (specific) box, named e.g., BOX–1, includes the slot A–KIND–OF with value BOX, the slot MATERIAL with value WOOD and the slot PURPOSE with value STORAGE. PURPOSE itself has the slot A–KIND–OF with value FUNCTIONAL–PROPERTY and a slot IMPORTANCE with value VERY–HIGH and so is included in the transfer frame. That is, the PURPOSE of a BOX, namely STORAGE, is "globally important" and is most likely salient for a metaphoric use of the nominal BOX1.

Winston restricts his scope to simple nominal metaphor in the physical domain. His principles, however, represent specific subjective (i.e., as viewed by humans) aspects of vehicle meanings that will be seen in Chapter 6 as a clear guide to selection of salient properties of vehicles for transfer to other domains.

The interpreted examples that **Weiner** [52] describes (similes rather than metaphors, but treatable in a similar way for our purposes) are restricted to those with only physical–domain predicates. Weiner adapts Ortony's [36, 37] theory of the role of salience and asymmetry in human metaphor processing–that is, that nominal metaphor depends on predicates which are of high salience for the vehicle nominal and low salience for the topic nominal. Weiner integrates salience information into hierarchically organized knowledge representations, and includes prototypicality relative to other properties or instances of the concept. A salient predicate is a predicate that indicates inherent prominence (great size, for example) or distance from other concepts in the hierarchy (cf. Winston's "extreme value" and "distinction with respect to the class of the object). The hierarchy distinguishes concept properties that are particular to the concept (and probably more critical to the metaphor) from those simply due to its place in the hierarchy. Also, the hierarchy makes it easy to determine the incongruity which is often associated with a good metaphor. The phrase, "Penguins are (like) wolves," for example, is more metaphoric than "Dogs are (like) wolves," because "dogs" and "wolves" share a superordinate, namely, "canines."

Figure 2.1 shows a representation of the Concept “apple.”

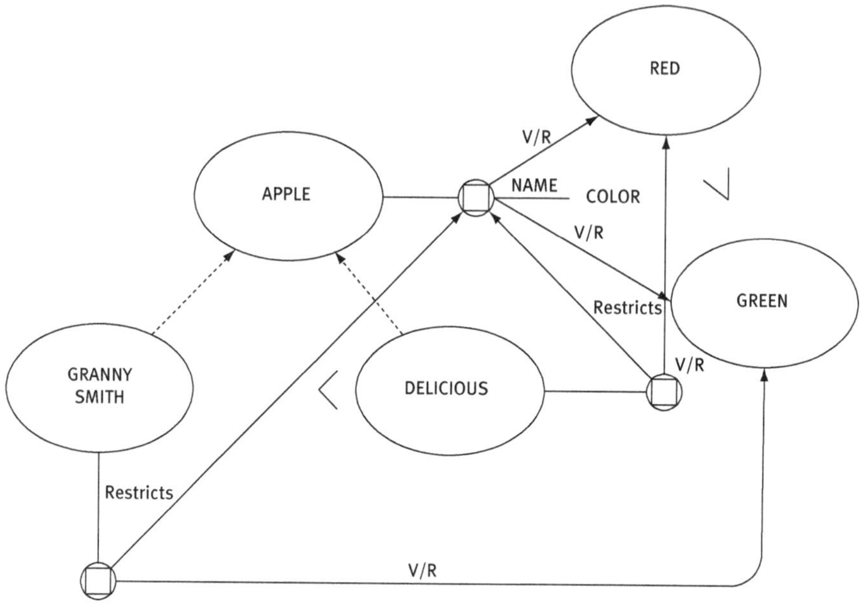

Note. From “A Knowledge Representation Approach to Understanding Metaphors” by E. J. Weiner, 1984, *Computational Linguistics, 10,* p. 11. Copyright 1984 by *Association for Computational Linguistics.* Reprinted by permission.

Fig. 2.1: Weiner’s representation of the nominal “apple.”

Prototypicality for apples is indicated by the “greater than” symbol, “>.” Thus the Value/Restriction, or V/R, on the Role COLOR, i.e. RED, is ranked as more typical than GREEN, and the Concept DELICIOUS (an apple type lower in the hierarchy than that of APPLE) is ranked as more typical than GRANNY SMITH, where DELICIOUS and GRANNY SMITH impose restrictions on the COLOR of APPLE. The properties RED and DELICIOUS thus serve as a prototype for processing the simile “Jane’s cheeks are like apples.”

Aside from tools to represent prototypicality, Weiner adds a *value range* as a potential value restriction and a *salience measure* as a role parameter. These specifications aid in the interpretation of “John’s hands are (like) ice.” “Hand” has a TEMPERATURE value range of 3 to 6, and the TEMPERATURE of “ice” has a (colder) value of 7 and a salience value of 1, where the possible salience range is from 0 to 1. Since JOHN’S HANDS does not restrict the Roles of the Generic Concept HAND, the expression can then be identified as (a) a hyperbole, since the V/R for the Role TEMPERATURE is extreme in the vehicle but not in the topic and (b) a metaphor, as the vehicle and the topic do not have a close shared superordinate in the taxonomy. Raising the salience of the TEMPERATURE Roles of the topic therefore gives the information that John's hands are extremely cold relative to normal hand temperatures.

The scope of Weiner's implementation is unclear. However, she has formally incorporated prototypicality and salience measure into an organized knowledge base as a support for interpretation of–at least–one type of physical–domain metaphor. Also, some aspects of this procedure can be used to determine metaphoricity and the "goodness" of a metaphor. Here one is reminded of the apparent contrast between "aptness" and acceptability of a metaphor noted earlier with respect to Fass's method, referring to Tourangeau and Sternberg [46]. It seems that the distinction between "great" metaphors and incomprehensible ones may sometimes be a fine one.

Veale and Keane [48] consider both verbal and nominal metaphor, and have implemented their system as applied to texts having to do with computer products. Their work is presented as "conceptual scaffolding," by which they mean that a very general metaphor interpretation in terms of associations between concepts serves as support for elaboration on the interpretation. The examples and interpretations they consider are based on Lakovian spatial conceptual metaphors; the "operators" they define for representation purposes are based on collocation, containment and experience.

Many elements of the scaffolding correspond closely with the abstract representation components described in Chapter 5. (For example, their basic "association" corresponds with the relation AT and their ATTEMPT component with TRY). The same can be said of their combinations of components, which has some linguistic/semantic consensus. However, Veale and Keane helpfully produce graphic illustrations of their components and examples, including applications of the *UP/DOWN* conceptual metaphor. Also, they provide some context for a more specific meaning of some verbs through narrow inheritance relationships (a laptop is a computer is a product). Considering the metaphoric use of the verb "kill," for instance, they note that "kill the PC" has differing meanings, depending on whether IBM or a virus is the subject/actor. In the former case, a product is "killed;" in the latter, a (specific) laptop/computer.

Some of what Veale and Keane present as elaboration has been captured in conceptual graphs in other systems. Given the sentence, "Ronnie gave Bonzo a bath," an association between "Bonzo" and "bath" through the elaborative "functional attribute transfer" of "Clean" provides the inference that Bonzo becomes clean, and moreover is "caused" by Ronnie. Here the function of "bathe" could be defined as "cause [object] to be [more] clean" directly. However, with their formalization of the concept of "opposite," their system can also infer that "Bonzo" was not clean before the transfer.

A more "real-world" metaphoric example given is "The PS/2 poisoned the PC market." Here an association between Poison and Market is created, which causes a transfer of Pain, with the result Down(PC.market) as Down(Physical-Stimulus) ->Pain. Along with this connection, the association Market = Animal is made.

As an example of "comparative (nominal) metaphor," Veale and Keane consider "The LX laptop is a real Godzilla." As salient properties they define extreme (cf. Winston [55]) attributes of the vehicle that also serve as functional attributes of the topic. Thus the functional attributes Weight, Speed and Size of the topic, "Laptop," are highlighted by the Heavy, Slow and Huge attributes of the vehicle, "Godzilla."

There does not seem to be any theoretical grounding of Veale and Keane's characterization of their system as two-stage, i.e., involving "scaffolding," though in practice the distinction is not without value.

2.2.2 Cross-domain metaphor

Kilpatrick [31] is one of the relatively few researchers who analyze nominal metaphor for which the topic is not purely physical and the predicates of either nominal are not necessarily simple physical properties. He directly determines which properties of the vehicle in nominal metaphor are salient. Like Winston, he views metaphor comprehension as "the ability to use shared stereotypic values to understand the metaphorical utterance" (p. 84). His method, however, is to store stereotypical and prototypical properties directly with the vehicle concept, as opposed to identifying possibly salient properties according to principles or a knowledge representation of the concept. Kilpatrick defines vehicle frames in terms of "stereotypic bundles" as well as nonstereotypic properties.

Thus for the example "The hearings are a blunderbuss," the stereotypic bundle for "blunderbuss" is retrieved from its frame definition. The stereotypic bundle (p. 85) consists of:

Function: kill object of aim
Accuracy: -7
Shot pattern: random, scattered

Kilpatrick's focus on stereotypic and prototypic properties can be seen as attention to the role of cultural subjectivity in metaphor. However, Kilpatrick has not analyzed the stereotypic properties themselves in other than metaphoric terms.

For example, his representations do not reveal how “kill” is to be understood with respect to the effect on the object. He states only that “kill” is somehow “tempered” when applied to the hearings.

Stereotypic descriptors with affective properties, incorporated in a lexicon, are used by Veale [47] in his model of affective comprehension and generation. A purpose is enhanced information retrieval (IR), because IR traditionally has treated all language input in literal terms, while for speakers, metaphoric language has been a useful method of communication. Stereotypical adjectives were confirmed by collecting triple-word phrases with their frequencies from the Google database and submitting corresponding similes to Google. To examine the role of affective properties, he considers conjoined adjectives, such as “happy and healthy,” and carries out research to determine whether such phrases are associated with a positive or negative environment. An interesting application (“Metaphor Magnet”) expands affective queries to include common metaphors, further expanded to include typically associated qualities.

The research of **Way** [49, 50] is included under “nominal metaphor,” although she also considers examples using predicate adjectives, such as “The car is thirsty.” Her system is distinguished by her “dynamic type hierarchy,” i.e., the reorganizing of the hierarchy through a metaphor interpretation. In this reorganization, new categories and links may be created and others removed. In a metaphor, the topic and vehicle share a common dominating type/category. If there is no such type, one is created by the use of masks (filters) operating on conceptual graphs of the topic and vehicle. Way examines the nominal metaphor “Nixon is the submarine of world leaders” and arrives at an interpretation on the basis of the superordinate Things-which-behave-in-a-secret-or-hidden-manner.

A more detailed description of Way’s analysis (though still simplified here), can be given with regard to the syntactically different example, “The car is thirsty,” where “thirsty” can be processed as the nominal “thirst.” The conceptual graphs for “car” and “thirst” are compared. The graph for “thirst” basically represents that thirst causes an animal to consume a liquid. Way uses her hierarchy to expand this representation in that “thirst” is subordinate to “need,” which is in turn subordinate to “requirement.” The conceptual graph for “car” includes three requirements, for gas, water and oil. All are liquids, and liquids correspond with what is consumed by an animal as a result of thirst. Of the liquids, “gas” is chosen because of its frequent and energy-driven need. As “car” and ”animal” are both a kind of MOBILE-ENTITY, a node is specified with this category applied to thirst. Given this supertype, both cars and animals are viewed through a mask as “mobile entities that require liquid,” giving inanimate concepts animate properties.

The dynamic treatment of her type hierarchy is ambitious and differs in a novel way from that of other researchers who use hierarchies. However, it is subject to similar problems. Categories such as Things-which-behave-in-a-secret-or-hidden-manner and even "mobile entities that require liquid" seem particularly questionable. The sense of an "ad hoc" nature makes it unclear that this method can confirm any principles for metaphor research, especially when nonphysical properties are involved.

In summary, most of the described work is oriented toward conventional, even dead metaphors. The underlying assumption is that many metaphoric expressions can be analyzed in the same way, regardless of their position on the spectrum of "aliveness." Nonconventional metaphor can be treated at least in part on similar mechanisms. One might speculate, however, that Lakovian conceptual metaphors as engrained in our language do not play as great a role in *human* comprehension of metaphor.[7] These are questions to be explored.

As for specific theoretical underpinnings, the determination of properties salient for any metaphoric phrase and therefore transferrable from vehicle to topic is a central concern for computational processing in terms of semantic analysis. In fact, Honeck, Voegtle, Dorfmueller and Hoffman [27] characterize metaphor and proverbs as "a *means* of focusing on salient aspects of complex events" (p. 132, italics mine). Several researchers specify various aspects of the task of determining salient properties in general. In particular, Winston states principles for the determination of salient properties; Weiner, following Ortony, integrates prototypicality and salience into her knowledge representations. Carbonell heuristically ranks types of properties as to likely salience.

In addition, for the quixotic goal of identifying all factors extended in metaphoric usage, attention to connotations, which are sometimes the "point" of a metaphor and may serve the purpose of its user, is an important but often neglected component of the analysis.[8] The cited computational work, with the exception of that of Narayanan, Dyer, the more recent system of Veale, and a reference to emotion by Indurkhya, does not explicitly distinguish experiential or subjective factors in their approaches to metaphor interpretation. Such

---

7 A difference in human comprehension has been noted in psychological experiments, [45] which suggest that humans no longer refer to mappings from spatial schemas in understanding temporal metaphors when these mappings become well-established; rather, they refer to an abstract relation that "covers" both domains.

8 Osgood [38] distinguishes two classes of metaphor by whether they are (only) "denotative" or "affective."

factors, along with linguistic considerations often disregarded in computational approaches to (especially) novel metaphor, play a role in the following chapters.

References

[1] *Cosmopolitan* **216 (3)** (1994). USA edition

[2] Agerri, R., Barnden, J., Lee, M., Wallington, A.: Invariant mappings and contexts in a computational approach to metaphor interpretation. In: R. Sanal, H. Mehta and R.K. Bagga (eds.), *Proceedings of the IJCAI–2007 Workshop on Modelling and Representation in Computational Semantics*, MRCS, Hyderabad, India, pp. 57–61. Morgan Kaufmann, San Francisco, CA (2007)

[3] Barnden, J., Glasbey, S., Lee, M., Wallington, A.M.: Domain-transcending mappings in a system for metaphorical reasoning. In: *Proceedings of the European Association for Computational Linguistics*, Budapest, Hungary, April 12-17, pp. 57–61 (2003)

[4] Barnden, J., Glasbey, S., Lee, M., Wallington, A.M.: Varieties and directions of inter-domain influence in metaphor. *Metaphor and Symbol* **19**, 1–30 (2004)

[5] Barnden, J., Lee, M.: An implemented context system that combines belief reasoning, metaphor-based reasoning and uncertainty handling, pp. 28–41. Springer, Heidelberg (1999)

[6] Carbonell, J.: Metaphor: A key to extensible semantic analysis. In: N.K. Sondheim (ed.) *Proceedings of the 18th Meeting of the Association for Computational Linguistics*, Stroudsburg, PA, June 19–22, pp. 17–21. Association for Computational Linguistics (1980)

[7] Carbonell, J.: Invariance hierarchies in metaphor interpretation. In: *Proceedings of the Third Annual Conference of the Cognitive Science Society*, pp. 292–295. Lawrence Erlbaum Associates, Hillsdale, NJ (1981)

[8] Carbonell, J.: Metaphor: An inescapable phenomenon in natural-language comprehension. In: W. Lehnert, M. Ringle (eds.) *Stragegies for Natural Language Processing*, pp. 415–434. Lawrence Erlbaum, Hillsdale, NJ (1982)

[9] Carbonell, J., Minton, S.: *Metaphor and Common-Sense Reasoning*, Rep. No. CMU-CS-83-110. Carnegie-Mellon University, Pittsburgh, PA (1983)

[10] Dyer, M.: Affect processing for narratives. In: *Proceedings of the National Conference on Artificial Intelligence*, pp. 265–268 (1982)

[11] Dyer, M.: Comprehension and acquisition of figurative expressions with phrasal/lexical memory. *Metaphor and Symbolic Activity* **4**, 173–201 (1989)

[12] Dyer, M., Zernik, U.: Encoding and acquiring meanings for figurative phrases. In: *24th Annual Meeting of the Association for Computational Linguistics–ACL*, July 10–13. Association for Computational Linguistics, New York, NY (1986)

[13] Falkenhainer, B., Forbus, K.D., Gentner, D.: The structure–mapping engine: Algorithm and examples. *Artificial Intelligence* **41**, 1–63 (1990)

[14] Fass, D.: *Processing Metonymy and Metaphor*. Ablex, Greenwich, CT (1997)

[15] Fass, D., Wilks, Y.: Preference semantics, ill-formedness, and metaphor. *American Journal of Computational Linguistics* **9**, 178–187 (1983)

[16] Fauconnier, G., Turner, M.: Blending as a central process of grammar. In: A. Goldberg (ed.) *Conceptual Discourse, Structure, and Language*, pp. 113–130. CSLI, Stanford, CA (1996)

[17] Fauconnier, G., Turner, M.: Conceptual integration networks. *Cognitive Science* **22**, 133–187 (1998)
[18] Fillmore, C.: The case for case. In: E. Bach, R. Harms (eds.) *Universals in Linguistic Theory*, pp. 1–88. Holt, Rinehart and Winston, New York (1968)
[19] Gentner, D.: Studies of metaphor and complex analogies: A structure-mapping theory. In R. Hoffman (Chair), *Metaphor as Process*, Symposium conducted at the annual meeting of the American Psychological Association, Montreal (1980)
[20] Gentner, D.: Are scientific analogies metaphors? In: D. Miall (ed.) *Metaphor: Problems and Perspectives*, pp. 106–132. Harvester, Brighton, England (1982)
[21] Grady, J., Oakley, T., Coulson, S.: Blending and metaphor. In: G. Steen and R. Gibbs (eds.) *Metaphor in Cognitive Linguistics*, pp. 234–246. John Benjamins, Philadelphia, PA (1999)
[22] Hayes, E., Bayer, S.: Metaphoric generalization through sort coercion. In: *Proceedings of the 29th Annual Meeting of the Association for Computational Linguistics (ACL-91)*, pp. 222–228. Berkeley, CA (1991)
[23] Hobbs, J.: *Metaphor, Schemata and Selective Inferencing*. Report No. 204 (1979) Stanford Research Institute, Menlo Park, CA
[24] Hobbs, J.: Metaphor interpretation as selective inferencing. In: *Proceedings of the 7th International Joint Conference on Artificial Intelligence*, Vol. 1, pp. 85–91. University of British Columbia, Vancouver, BC, Canada (1981)
[25] Hobbs, J.: *Metaphor and Abduction*. url http://www.dtic.mil/tr/fulltext/u2/a259421.pdf (1991). Technical Report 508
[26] Hobbs, J.: Metaphor and abduction. In: A. Ortony, J. Slack, O. Stock (eds.) *Communication from an Artificial Intelligence Perspective: Theoretical and Applied Issues*, pp. 35–58. Springer, Berlin (1992)
[27] Honeck, R., Voegtle, K., Dorfmueller, M., Hoffman, R.: Proverbs, meaning, and group structure. In: R. Honeck and R. Hoffman (eds.) *Cognition and Figurative Language*, pp. 127–162. Lawrence Erlbaum Associates, Hillsdale, NJ (1980)
[28] Huang, X., Huang, H., Liao, B., Xu, C.: An ontology-based approach to metaphor cognitive computation. *Minds and Machines* **23**, 105–121 (2013)
[29] Indurkhya, B.: *A Computational Model of Metaphor Comprehension and Analogical Reasoning*. Ph.D. Thesis, Report No. 85/001 (1985)
[30] Iverson, E., Helmreich, S.: Metallel: An integrated approach to non-literal phrase interpretation. *Computational Intelligence*, Special Issue on Non-Literal Language **8**, 477–493 (1992)
[31] Kilpatrick, P.: An A-frame model for metaphor. In: *Proceedings of the International Conference on Cybernetics and Society*, pp. 83–87 (1982)
[32] Lakoff, G., Johnson, M.: *Metaphors We Live by*. Chicago University Press, Chicago, IL (1980)
[33] Lytinen, S., Burridge, R., Kirtner, J.: The role of literal meaning in the comprehension of non-literal constructions. *Computational Intelligence* **8**, 416–432 (1992)
[34] Martin, J.: *A Computational Model of Metaphor Interpretation*. Academic Press, New York (1990)
[35] Narayanan, S.: Moving right along: A computational model of metaphoric reasoning about events. In: AAAI (1999)
[36] Ortony, A.: Beyond literal similarity. *Psychological Review* **86**, 161–180 (1979)
[37] Ortony, A.: The role of similarity in similes and metaphors. In: A. Ortony (ed.) *Metaphor and Thought*, pp. 186–201. Cambridge University Press, New York, NY (1979)

[38] Osgood, C.: The cognitive dynamics of synesthesia and metaphor. In: R. Honeck, R. Hoffman (eds.) *Cognition and Figurative Language*, pp. 203–238. Lawrence Erlbaum, Hillsdale, NJ (1980)

[39] Russell, S.W.: Information and experience in metaphor: A perspective from computer analysis. *Metaphor and Symbolic Activity* **1**, 227–270 (1986)

[40] Salmans, S., DeFrank, T., Buresh, B., Hubbard, H.: A ford in high gear. *Newsweek* **20**, 13 (1975)

[41] Schank, R.: *Conceptual Information Processing*. North Holland, Amsterdam (1975)

[42] Schank, R., Abelson, R.: *Scripts, Plans, Goals and Understanding*. Lawrence Erlbaum, Hillsdale, NJ (1977)

[43] Shutova, E.: Automatic metaphor interpretation as a paraphrasing task. url https://www.researchgate.net/publication/220817544_Automatic_Metaphor_Interpetation_as_a_Paraphrasing_Task (2010). Accessed 23-5-15

[44] Suwa, M., Motoda, H.: Learning metaphorical relationships between concepts based on semantic representation using abstract primitives. In: Fass, D., Martin, J., Hinkelman, E. (Eds.), *Proceedings of the 12th IJCAI Workshop on Computational Approaches to Nonliteral Language: Metaphor, metonymy, idiom, speech acts and implicature* IJCAI-CANL, August 24–30, pp. 123–131. Tech. Report No. CU-CS-550-91. University of Colorado at Boulder, Boulder, CO (1991)

[45] Teuscher, U., McQuire, M., Collins, J., Coulson, S.: Congruity effects in time and space: Behavioral and erp measures. *Cognitive Science* **32**, 563–578 (2008)

[46] Tourangeau, R., Sternberg, R.: Understanding and appreciating metaphors. *Cognition* **11**, 203–244 (1982)

[47] Veale, T.: Once more, with *feeling*! using creative affective metaphors to express information needs. url http://www.computationalcreativity.net/iccc2013/download/iccc2013-veale-1.pdf (2013). Accessed 22-05-15

[48] Veale, T., Keane, M.: Conceptual scaffolding: A spatially–founded meaning representation for metaphor comprehension. *Computational Intelligence*, Special Issue on Non-Literal Language **8**, 494–519 (1992)

[49] Way, E.C.: *Knowledge Representation and Metaphor*. Kluwer Academic Publishers, Dordrecht, The Netherlands (1991)

[50] Way, E.C.: Metaphor as a mechanism for reorganizing the type hierarchy. *Journal of Knowledge-Based Systems* **5**, 223–232 (1992)

[51] Weber, S.: A connectionist model of literal and figurative adjective noun combinations. In: D. Fass, E. Hinkelman, J. Martin (eds.) *Proceedings of the IJCAI Workshop on Computational Approaches to Non–Literal Language: Metaphor, Metonymy, Idiom, Speech Acts, Implicature*, Sydney, Australia, August 24–30, pp. 151–160 (1991)

[52] Weiner, E.J.: A knowledge representation approach to understanding metaphors. *Computational Linguistics* **10**, 1–14 (1984)

[53] Wilks, Y.: Knowledge structures and language boundaries. In: IJCAI, pp. 151–157 (1977)

[54] Wilks, Y.: Making preferences more active. *Artificial Intelligence* **11**, 197–223 (1978)

[55] Winston, P.: Learning by creating and justifying transfer frames. *Artificial Intelligence* **10**, 147–172 (1978)

3 A Semantic–Component–Based Approach

There are several dimensions according to which we can view the field of linguistic metaphor. We can consider a single sentence vs. a larger piece of discourse; given the former, a single sentence in which only a part is metaphoric (part–sentence or partial metaphor) vs. one which is totally metaphoric; a metaphor which is all within the physical domain vs. one which crosses domains; conventional vs. unconventional (novel) metaphor; and metaphor based on a recognizable similarity between vehicle and topic vs. metaphor in which the originator *creates* a similarity. While the latter may roughly correspond to conventional vs. novel metaphors, the line between them is fuzzy, and there are many degrees of novelty.

In terms of these dimensions, the main topic of our discussion and object of the computational task is part-sentence metaphor, where a sentence consists of a simple predication, with particular attention to the interesting challenge of cross–modal (or cross–domain) metaphor, as detailed in later chapters. As the described computer program handles both conventional and nonconventional metaphor in the same way, the conventional–unconventional distinction does not play a role in the computational approach presented here. The similarity–creating vs. similarity–recognizing distinction is interesting theoretically in itself, and is left to later discussion. The program grounded in this exploration of the semantics of metaphor processes part-sentence metaphor containing either metaphorically used verbs or metaphorically used nouns, with discussion of within–domain metaphorical use of verbs restricted to the present chapter. In order to focus on the semantics of metaphor, the syntax of our target examples will be kept as simple as possible; even such grammatical elements as determiners and quantifiers, interesting though they may be from a logical perspective, are ignored.

3.1 Types and examples

Part–sentence or partial metaphor distinguishes itself from metaphor in which a whole sentence refers to a situation intended to be analogous to another situation.

Proverbs, which have been studied by Honeck et al. [13], are examples of whole–sentence metaphor, e.g., “Don’t count your chickens before they are hatched,” “Haste makes waste,” etc. While a proverb could be re-phrased as a single general, literal sentence, it may apply by analogy to many specific situations.

Whole paragraphs or stories may be metaphoric, the latter called allegories. The following paragraph example [12], though phrased as a simile, not only illustrates metaphoric discourse, mixed with a few phrases that apply to both topic and vehicle, but also shows how different metaphors can convey two different characterizations:

> Philosophy, at its rigorous best, is a bit like rock-climbing: you slowly make your way up a craggy conceptual cliff, gaining a foothold here and there by dint of hard logic, all the while trying to keep your head straight as the oxygen thins.
>
> For Frayn, philosophy is more like windsurfing. He scuds from one frothy wave to the next, leaving a spray of question marks in his wake. (p. 14)

The following paragraph (though rather amusingly mixing metaphor with literal language) shows how the use of metaphoric language can seek to persuade. [7]

> This is not the fox guarding the hen house; this is the fox guarding the hen house while selling synthetic derivatives whose value increases with every hen he gobbles up, and who burns down the hen house so he can collect on his fire insurance policy, and then gets the government to build him a new hen house at taxpayer expense. And 'then,' after that, he still gets to guard the new hen house.

Some annotated examples of within-(physical) domain metaphor follow. Complex original examples are reduced to a simple predication. As in the remainder of this chapter, examples include both "live" and "dead" metaphors, since they are treated equally, e.g., without reliance on polysemic lexicon definitions.

"...But in the middle of that night Aeneas plowed the coastal sea." [26] ("ship plow sea") "Aeneas" metonymically replaces "his ship."

"...squeezing money out of the wasteful system and freeing it up for more productive uses" [2] ("squeeze money") "Money" is not truly physical in this sense.

"l'avalanche de cette 'neige' mortelle"[1] Translation: "avalanche of this deadly 'snow'" ("drug is snow") "Avalanche" extends the metaphoric language with a "wide use is avalanche" metaphor. This metaphor as used here can be considered to be cross-domain, since "avalanche" applies to an action (usage) rather than to a physical mass. However, in another context it could *conceivably* be the physical-domain metaphor "(falling) drug is avalanche," as a drug powder pouring out, e.g., from a large bag.

1 Journal de Geneve, Feb. 3, 1982, article no longer available.

"The weather is spoiling"–a partly physical–domain metaphor, since weather is physical but has an action component. This is an evaluative statement; one could as well say objectively (without evaluation), "the weather is (becoming) stormy."

"...among the passages were to be found single steps 'that seemed like spies from lost battalions, lying in wait and wondering where the rest had gone'..." [25] Arguably an amusingly overdone metaphor. Reducing this metaphor to "steps are spies" would miss the flight of imagination.

"Every four years, the NH primary showers the campus with candidates."[2] (the verbal metaphors "primary showers campus, primary showers candidates (as in "it's raining cats and dogs"); the nominal metaphor "candidates are showers;" the metaphoric noun compound[3] "candidate showers")

The documents percolated up from the bottom of the pile.

The highway snakes...

The highway is a snake

Sunlight was filtering through the curtains.

"See Dick and Jane vacation to Hawaii. See them take in the volcano. See them get close to the edge. See the volcano take in Dick and Jane." [9] ("humans take–in volcano" vs. the literal "volcano take–in human") Both cross–domain metaphor and literal usage respectively, resulting in a kind of pun.

Pinocchio lizard. ("proboscis is [like] Pinocchio's nose")

Idea factory

France is home plate (by child looking at map of Europe).

Billboards sprout...

Billboards are warts (on the landscape)

The rug drank the wine.

Mountains undulate...

These examples, with the possible exception of the "weather" and "money" phrases, can be contrasted with cross–modal metaphors, where the difference in domain between topic and vehicle is a difference between "modes" or

2 UNH "Campus Journal," out of print
3 Also called "compound nouns."

“conceptual domains,” which are very broad (and roughly Aristotelian), corresponding to human faculties. While the domain difference in “Billboards sprout...” is between an artificial object and a natural object (plant), both physical, the difference in mode or conceptual domain between “music” and “flooded” in “Music flooded the room” is between that of physical concepts (“room”) and that of a sensory/sound modality (“music”). Similarly, in “The workshop sprouted ideas,” “sprouted” is physical, while “ideas” (and the metaphoric sense of “sprout”) is in a mental domain.

Some metaphors are anthropomorphisms or personifications, in which human/animate features are ascribed to an inanimate actor. Examples are “The trees began to think about dropping their leaves,” the example of Fass and Wilks, [6] “My car drinks gasoline” and “The sun smiled on me.”

3.2 Part-sentence metaphor

Part-sentence input to an early [21] experimental program contained a literally or metaphorically used verb. The syntactic structure of the metaphors had the simple syntactic format either “subject (noun) verb direct-object (noun)” or “object (noun) (intransitive) verb.”

As prepositions in combination with verbs are often the main contributors to the meaning of a verb structure, references to “verbs” include expressions consisting of a verb followed by the preposition it “goes with,” i.e., “phrasal verbs.” In the program's input, phrasal verbs are hyphenated, as in “plow-through.” Support for using phrasal verbs as a unit (simply called verbs in this discussion) is illustrated by the near-equivalence of two verbal concepts, one lexicalized by simple verb and the other by a verb followed by a preposition, such as “inhabit” and “live in.”

The preposition says much about the relationship between the nominals linked by the verb. For the phrase “leap to,” for example, it is the preposition “to” which connects the “leaper” to the goal. In a comparison of the verbs “plow” and “plow through,” it is the preposition “through” which conveys the basic interpretation component of motion; the verb “plow” by itself emphasizes what is done to its object. A thrown vase may plow through a window; “plow” would not make sense here. The lexicon used therefore highlights different components for, e.g., the verbs “plow” and “plow-through.” (See Brugman and Lakoff [3] for a more comprehensive view of metaphoric meanings of prepositions.)

Metaphoric sentences were either within-(physical) domain or cross-domain. Conventional/mundane and unconventional/novel metaphor were treated the same way, reflecting the fact that much language not even noticed as metaphoric has a metaphoric origin.[4] Pseudo-paraphrases considered to be literal were generated from this input via an "abstract"-component-based representation.

The remainder of this chapter describes some of the basic concepts, further developed in succeeding chapters, used in the metaphor analysis of part-sentence metaphor. The principal components are domains and structures and their relationship, considered from a general linguistic and specifically semantic/conceptual perspective.

3.2.1 Linguistic evidence

In the area of linguistics, Jackendoff [16, 15] credits Gruber [8] in noting that the spatial concepts of location and motion have expression in other "fields" (domains), i.e., generalize across modalities. Jackendoff has formulated this observation, since made by other linguists and metaphor researchers, as the Thematic Relation Hypothesis, which he claims has an important role in cognition.

As examples, he gives the analogically related pairs:

The train traveled from Detroit to NY

GO (THE TRAIN, DETROIT, NY)

POSITION

Harry gave the book to the library

GO (THE BOOK, HARRY, THE LIBRARY)

POSSESSION

Jackendoff has laid out a comprehensive taxonomy of spatial relationships from a linguistic perspective and has copiously illustrated the extension of spatial to temporal concepts with examples, including:

The meeting is at 6:00.
We moved the meeting from Tuesday to Thursday.
Despite the weather, we kept the meeting at 6:00.

4 see Lakoff and Johnson [18] for demonstration of the ubiquity of conventional metaphoric language.

The statue is in the park
We moved the statue from the park to the zoo.
Despite the weather, we kept the statue on its pedestal.

Further examples of space–to–time extension, on the basis of Clark's [4] observation that the observer rather than the temporal concept can serve as the reference point are:

Tuesday crept by
Christmas is fast approaching
Our future lies ahead of us

The freight train crept by
The tiger is fast approaching
The frontier lies ahead of us

These show metaphoric extension, though with dead metaphors.[5] Lakoff and Johnson [18] characterize such extensions as based on a conceptual metaphor, and in the NLP field, Hayes and Bayer [10] as described in Chapter 2, have concentrated their metaphor research on implementation of space–to–time extension.

3.2.2 Domains

As stated earlier, there are both similarities and differences between a literal and a metaphoric sense of verb usage. This fact plays a role in both recognition of verbal metaphor and literal interpretation of its meaning. For literal language, the domain of the verb and that of the object must minimally be consistent (though not necessarily the same, as will be seen) for the phrase to make literal sense, and must *be made* consistent, to make metaphoric sense.

Domain consistency for literal coherence particularly makes sense when we consider that if a verb represents an action upon an object, the meaning of the object can be thought of as included in the meaning of the verb. For example, the verb "to hoard" always implies an object which is hoarded. When this consistency does not hold, we have either a potential metaphor ("He hoarded compliments") or an incoherent phrase. In the case of a "subject verb direct–object" input, domain tests to recognize metaphor were between verb and direct object; in case of "subject verb" input, between subject and verb. The inconsistency of a metaphor is resolved

5 Jackendoff, however, distinguishes such "thematic structuring" from metaphoric structuring, because of the variety, additional purposes and "artistic" nature of metaphor.

when the verbal concept is comprehended as extended to the domain of the object nominal; the extended part then represents the similarity.[6] A comprehensibly extended verb can be seen as analogous to the verb in its literal sense, structural elements being the same but the original (vehicle) domain of the verb different.

For verbal metaphor, then, we must have a means of assigning domains to verbs and nominals, requiring a domain taxonomy. The domain taxonomies used to determine the metaphoric nature of a phrase differ for within-(physical)-domain and cross–domain metaphor in terms of granularity. In contrast to the domains or modes of cross–domain metaphor, physical-domain metaphor resides within one mode, i.e., relying on extension between two physical subdomains. Thus in the frozen metaphoric usage, "The ship plowed through the sea" or "The ship plowed the waves," the verb "plow" is extended from the domain of farming and land domain to that of sailing or the sea. Both domains are physical, but one cannot literally "plow" the sea in the *original* sense of "plow."

The proposed domain taxonomy comprises four "conceptual domains" of verbal and nominal concepts[7] plus two, SPACE and TIME, which have not yet been used in the program. Conceptual domains are so called to distinguish them from detailed domains that are all within the physical conceptual domain. Each of the four domains has one additional level of subdomains, described in Chapter 5, and are defined as follows. (Words used as examples here and throughout the book are used in just one common sense; particular word senses would be submitted to the metaphor procedure by the "host" NLU program).

PHYSICAL, e.g., "jump"
MENTAL, e.g., "forget"
SENSORY, e.g., "listen"
CONTROL, e.g., "bequeath"

The additional SPACE and TIME domains are illustrated by "room" and "day" respectively.

That three of these domains correspond to the "primitive ACTs" PTRANS, MTRANS and ATRANS of Schank [22], is not surprising. It is evident that these ACTs represent the (structure of) various kinds of transition in the PHYSICAL, MENTAL and CONTROL (possession) domains respectively. Teasing out these domains reveals a perceived analogy of conceptual structure in the various domains.

---

6 The nominal concept may instead be extended to the domain of the verbal (see Levin [19] for comparisons of the two kinds of extension), but the number of such cases is relatively small.
7 In defining these domains I benefited from discussion with Larry Tesler.

The PHYSICAL domain is the domain of concepts which are either wholly spatial/material or have components in that domain. Spatial relationships as visible phenomena are prolific in structuring expressions in other domains.

The MENTAL domain is the domain of active or stored mental concepts, e.g., thought and memories. The "object" of a MENTAL verb is a mental representation, as in "She considered *the idea of selling seashells.*"

The SENSORY domain, as perception, serves as a bridge between the PHYSICAL and MENTAL domains. While domain consistency in many cases means that verb and object have the *same* domain (one sees images, hears sounds, tastes flavors, smells vapors, feels pain), in language these "transmitters" are usually skipped, so that, e.g., seeing or tasting food can be considered domain-consistent.

The CONTROL domain involves relationships or possession, power, obligations, etc.[8] These are sometimes expressed through modal auxiliaries, such as "can," "may" or "must." Possession as a CONTROL relationship is distinguished from purely PHYSICAL or location relationships. The verbs "have" and "give" (a physical object), for example, are defined in terms of the CONTROL domain rather than the PHYSICAL domain.

Examples of extension from the literal domain of a verb to that of its relevant object are:

PHYSICAL to MENTAL

He closed his mind.
Protests rained on the government.
Europe and America are drifting apart.
Could also be literal.
Kohoutek's tail points to its origin.
("Tail" refers to "information about the comet's tail" and "origin" perhaps to "information about its origin;" could also be literal, with "points" a physical concept and "origin" a spatial co-ordinate.)

PHYSICAL to SENSORY The music/fragrance flooded the room.

PHYSICAL to CONTROL

The privilege of emptying the trash landed in his lap.
Includes an ironic use of "privilege."

8 Schank [?] has characterized a class of such relationships in terms of triangles instantiated, e.g., by an authority which resolves conflicts between two opponents.

Control of the situation sailed away.
Their rights took the train to the coast.

MENTAL to PHYSICAL The fudge didn't agree with me.

MENTAL to SENSORY His music is asking for some more work

MENTAL to CONTROL

Duty calls
Perks are beckoning

SENSORY to PHYSICAL Trees whistled in the wind.

SENSORY to MENTAL

I searched for an answer.
I'd like to x–ray his reasons.

SENSORY to CONTROL These requirements smack of discrimination.
(Passive form of "When I taste these requirements...")

CONTROL to MENTAL She donated an idea.

CONTROL to SENSORY She usurped his view.

CONTROL to PHYSICAL My feet are demanding a rest.

It can be seen that extensions to the PHYSICAL domain as shown are anthropomorphisms, as they come from animate faculties applied to inanimate concepts. They do not appear to be of the same kind as metaphoric extension to other domains, are not as prevalent and are not considered further here.

Some of the other above types of extension from nonPHYSICAL domains appear to be less easily illustrated, presumably because of the obvious immediacy of bodily experience in creating metaphoric usages, consistent with "embodied cognition" as put forth by Lakoff and Johnson [18]. The lack of extensions from the MENTAL domain, other than anthropomorphisms, may be due to the fact that MENTAL objects are indeed "all in the mind." The object of a MENTAL verb is always internal – an idea or other kinds of information, no matter what it represents.

Extensions from the SENSORY and CONTROL domains are, however, conceivable, if not nearly as frequent as from the PHYSICAL domain. SENSORY (especially visual) domain extensions are fairly prevalent, as the senses as receptors are a means of "taking in" information and, in people, operate in conjunction with the MENTAL domain. Consider the prevalence of "I see," "It looks to me..." and "I hear you" to refer to understanding, a MENTAL concept.

Extensions from the CONTROL domain present a less obvious case. Domain consistency for CONTROL verbs, connected as they are to modal auxiliaries (can, must, may, etc.), means control over a physical object or an action. Usurping a right to do something is literal in the CONTROL domain, but usurping a view is metaphoric, as "view" is a SENSORY concept.[9]

With CONTROL verbs especially, it is sometimes not obvious what the true object of a verb is. For example, if we say that the CONTROL verbs "allow" and "prohibit" are used literally only when the OBJECT is an action (representing domain consistency for this domain), is "He allowed/prohibited slang" metaphoric? It does not seem so. Rather the action "*use* of slang" is omitted. The role of ellipses and implicit actions in language is evidently relevant to an determination of what is metaphoric, metonymic or incoherent. Given the relative infrequence and complexity of the extensions from nonPHYSICAL comains, then, these domains *as vehicles* will not receive as much attention in this book, without being neglected. Extensions from the PHYSICAL domain, which provides the most prevalent, experiential and rich source of metaphor will serve as the main substance of the discussion.

3.2.3 Structure extension

As pointed out in Chapter 2, Ortony [20] notes high salience of a feature of the vehicle–domain concept and low salience of that feature in the topic–domain concept as a mark of metaphor. He gives as an example "billboards are warts," where "billboards" is the topic and "warts" the vehicle; the feature in question is "ugly protrusions." While this principle does not apply to all metaphors, it is certainly operative for this example and for many metaphors of this format.

For verbal metaphor as discussed above, the object nominal is in the topic domain; the verbal concept in its literal sense is in the vehicle domain. Alternatively, one could denote the literal sense of the verb as the vehicle and its extended sense as the topic, which amounts to the preceding indication, since the extended sense is in the domain of the (topic) nominal. Salient features of the vehicle are what are transferred to the topic.

Ortony refers to salience of nominal concepts. What, then, is transferred from a *verbal* concept? Whatever the "point" or richness of a metaphor, its use must satisfy certain structural constraints. In a linguistic context, a noncontroversial

9 The frequent (nonmetaphoric) occurrence of a "throne" as an object of "usurp" in a historical context is metonymic.

candidate for the most basic structural component is causation, as many verbs express an act that has a result. (See Schank and Abelson [23] for a discussion in the context of other types of causation and inferences.) The *similarity* exhibited between the literal and metaphoric usage of a verbal concept involving causation tends to reside in the result or "effect" portion of a representation. In the following sets of examples, each of which contains a literal, within–domain metaphoric and cross–domain metaphoric sentence, we can see structural analogies between the effects within each set:

> A ten-pound baby was born.
> A three-star restaurant was born.
> Her belated ambition was born.

Result: The baby/restaurant/ambition exists.

> He shoveled more snow onto the six-foot peaks.
> He shoveled the oatmeal into his mouth.
> He shoveled up the praise.

Result: The snow/oatmeal/praise is in a different literal or metaphoric location.

> She embroidered the patch for the quilt.
> She embroidered the wall with graffiti.
> He embroidered his stories with fake incidents.

Result: The patch/wall/memories are added to.

The last example also carries the meaning of adding something positive (if we regard the "graffiti" example as ironic).

The following examples also share a common component, in being NON-causative:

> The beans simmered too long.
> Lava simmered beneath the crust.
> His resentment simmered too long.

Something progresses at a low level.

> The fish was sandwiched between the two slices.
> The basalt was sandwiched between the granite layers.
> The bad news was sandwiched between the good news.

A specific topology is shared.

As states can have temporal boundaries, as in "*stop* simmering," the beginning and end of a state are also basic, transferrable elements of structure. The extension of such components suggests that literal senses of verbs be represented in the lexicon in terms of CAUSEd (by an AGENT or event) or noncaused STATEs, along with a BEGIN or END component (with the redundancy that a caused state can always be viewed as a begun state). These elements, along with the conceptual domain in which a verb is considered literal, are minimal components for the determination of metaphoric verb extensibility.

3.3 Within-domain metaphor

The way an expression is interpreted, if at all, depends on how concepts of the expression fit together. For the syntax we are considering, inconsistency between conceptual domains indicates the *possibility* of a cross–modal metaphor. For within–domain metaphor, however, the conceptual domains of both the verb and its object are PHYSICAL. Domain inconsistency still applies to the determination of metaphor, but there are many real–world PHYSICAL domains. How do we know, then, whether a phrase of our considered format containing a verb and object both in the PHYSICAL domain is metaphoric?

3.3.1 Feature-based constraints

In the context of a host NLU program, a test for metaphor could be triggered by the recognition that a tentative construct is incoherent in a literal sense. For the metaphoric war horse, "The ship plowed through/plowed the sea" the domain of land differs from the domain of the sea, so the expression is not literally coherent, although since the metaphoric use of "plow" and "plow through" are quite conventional, the metaphoric analysis could be skipped in favor of retrieving the frozen sense from the lexicon. A more novel PHYSICAL–domain metaphor is, "they were sardined into a strip of shade behind the house." [27] Here it is first necessary to capture the making of the nominal "sardine" into a verb, which would come from world knowledge that sardines are customarily packed tightly into cans, i.e., contiguous in a can. Sardines and people are in different PHYSICAL domains, so the example is not literal.

As these examples (the former arguably) are not literally comprehensible, what makes them metaphorically comprehensible, while, e.g., "they were turtled into a strip of shade" is not? As concerns disambiguation, how do we know that in the example, "The coffee–pot and the toys plowed through by the toddler...," "the coffee–pot" was probably not "plowed through"?

The situation is similar for the format <subject> <verb>. The example, "Flowers cascaded (down the hill)" can be considered metaphoric, since it is comprehensible (why?) and has a (PHYSICAL) domain difference. Some of these phrases are anthropomorphisms, such as Indurkhya's [14] "The sky is crying" or "The pond is smiling."

There are therefore constraints that must be satisfied in order to determine coherence; only objects of certain types can comprehensibly integrate with certain verbal concepts. Only water and "things similar in some respect," e.g., fluids, can literally cascade; only humans (and other animals in a *conceivable* sense) can literally cry.

How to define these types with clear borders is an obstacle in theory as well as in practice. We may, however, be able to roughly specify dependencies between verbal and nominal concepts in terms of not only literal selectional restrictions, but in terms of what is conceivable for a *metaphoric* interpretation. A natural language parser could benefit from such knowledge in the task of attaching words to their proper structures,[10] resolving anaphoric references, determining correct senses, and detecting nonmetaphoric tropes.

For example, "plowing through a coffee–pot" is difficult to conceive of, but "plowing through toys" is not, because the plural number of "toys" allows (relatively) unhindered transit, as does the ocean or anything fluid, whether liquid or granular. In addition, given that an object either has the function of "going" or is propelled, and the two objects involved have rough relative sizes, one object "plowing through" another is also conceivable.

Such conditions conjure up the possibility of many categories and subcategories. As Arnheim [1] states with respect to hierarchies of categories:

> Each individual thing would be explicitly assigned to as many groups as there are possible combinations of its attributes. A cat would be made to hold membership in the associations of material things, organic things, animals, mammals, felines, and so forth, all the way up to that exclusive club for which only this one cat would qualify. Not only this, but our cat would also belong among the black things, the furry things, the pets, the subjects of art and poetry, the Egyptian divinities, the customers of the meat and canning industries, the dream symbols, the consumers of oxygen...

A system consisting entirely of fixed categories, then, is not a good means of characterizing concepts and is difficult to revise. A representation system based on features, with a few high–level categories, by contrast, is flexible, allows

---

10 Resolution of this problem is similar to that pioneered by Katz and Fodor [17] in their attempt to determine correct syntactic structures through reference to semantic descriptions.

overlap, and can easily be revised. An initial conceptual feature set for nominals is presented briefly here, for the purpose of determining metaphoric coherence of PHYSICAL–domain phrases, with examples of concepts which do or do not (before or after the slash) satisfy a positive value for the feature. The set, though clearly not complete, is much smaller than a set of descriptors for all objects and their literal relationships would be, as conceivability rather than usualness is the criterion.

Most of the features reflect topological properties (see Tab. 3.1):

Tab. 3.1: Features with examples.

Feature	Examples
+/–	
PART	roof, step/house, proof
SHAPE	rainbow, idea/fog, geography
CONTAIN	shoe, soup/gold
FIXED	field, building/bird, ball
1-DIMENSIONAL	fence, streak/ball, flash
2-DIMENSIONAL	field, table/pole, statue
FLUID	plural concept, river, sand/desk, tree

Obviously these should be considered as idealizations of actual physical properties; anything can actually contain something, except the most minute particle, perhaps a gluon, and a drowning person will recognize that rivers are not truly one-dimensional. But for example, something can be "along" a river in a way that it cannot be "along " gold, a ball, etc. It is also reasonable to expect a further level of distinctions for some of these features. For example, there are different types of containment: an object can be within a hollow container or a bounded surface area or immersed in a fluid; an object can be contained in a solid.

HUMAN and ANIMATE, with HUMAN here implying ANIMATE, are included, as they are in most taxonomies. In addition, an object may be +/–DYNAMIC, such as "boy," "motor" or "story," but not "key" or "stone." This feature then refers not to whether something has moving parts, though it may, but to whether the concept has some kind of "continuous existence" by itself, other than mere spatial presence. Such concepts, sometimes thought of as being "alive," can for example be metaphorically killed.

SIZE, with a 0–5 scale reflecting human interaction with the concept (able to be held in hand, for instance) rather than absolute size, also determines certain relationships, though it is a pragmatic rather than a conceptual feature.

Finally, an important descriptor of a nominal is its function or typical action; however, this descriptor is mainly of use for nominal metaphor, discussed in Chapter 6.

A variable value for some of these features is also allowed for (a rubber tree can be FIXED or in a movable pot). However, even with this fall-back, the approximate nature of these features, or perhaps any features, often translates into uncertainty of the determination of the coherence of a verbal concept and its nominal object. Constraints of this type are related to more comprehensive work by other researchers on "common-sense reasoning" with regard to non-metaphoric language.

3.3.2 Interpretation

The experimental program which implemented the given approach to within-domain metaphor provided a *basis* for distinguishing metaphoric, literal and incomprehensible expressions with the simple criteria presented.

3.3.2.1 Metaphoric vs. literal

An expression may make sense literally, but also have metaphoric meanings. The example, "Europe and America are drifting apart," since "Europe" and "America" have both PHYSICAL and nonPHYSICAL components, may refer to either continental drift (within PHYSICAL-domain) or to differing thought, perceptions or ambitions (cross-domain). The concepts from which Europe and America are drifting are each other, both having components in all domains. A book also can be interpreted in a PHYSICAL and nonPHYSICAL (MENTAL) sense. It is usually but not always clear which sense is intended. The sentence "She abandoned the book" could refer to either leaving the PHYSICAL book somewhere or stopping work (writing or reading) on the MENTAL book. Such nominals can be treated in the lexicon either with dual senses or with a dual set of components. Some researchers (see Fass [5]) treat this dual sense of nominals such as "book" as metonymy, where one sense is used in place of the other.

Similarly, some verbs have become so "generic" that they are perceived as applying to all domains, or are seen as metaphoric by one lexicon editor and literal by another. The verb "destroy" is easily seen as applicable to domains other than the PHYSICAL. It is of little consequence which domain is assigned to such a verb in the lexicon. If the verb is used together with its lexicon domain, it will be interpreted as literal-otherwise as extended and therefore metaphoric, but the interpretations should have the same sense, if not the same level of detail.

3.3.2.2 Procedure

Distinguishing between PHYSICAL–domain metaphor, literal expressions and incoherent phrases was approached through the satisfaction of "broad," "narrow," and no constraints respectively, with the broad constraints corresponding to the initial feature set presented above. As the criteria and lexicon were relatively undeveloped, the program served mainly as a testing tool and initial indicator as what could be achieved, taking into consideration the variability of language, the world and human judgment. The bias should be to attempt to interpret an expression rather than to reject it, since the input submitted for metaphoric interpretation is a potential parse, and presumably more prevalent than misparses.

The program which implemented this method is described in [21]. Briefly, the semantic definitions of the verb and noun(s) were retrieved from the lexicon, giving feature–based information for both and the basic structure for the verb. A satisfaction of narrow restrictions indicated a literal interpretation. For example, the verb "drink" has as a narrow restriction that the drinker be +ANIMATE and as a broad *and* narrow restriction that the object be +FLUID.

Roles of the nominals are determined, depending on which syntactic form is input. For input with two nominals, this procedure includes a test and adjustment for the reversal of the two nominals, enabling the same interpretation for "prosperity leap to country" and "country leap to prosperity," i.e., "country start be prosperous."

Control is then passed to other routines, depending on which role configuration is present. The output is then the interpretation (or no interpretation if no criteria are satisfied). The interpretation was given in terms of pseudo–English phrases in order to show capabilities of the method. In an operational context, the output would be in the form of symbols used by the NLU program embedding the metaphor processing capability.

Examples which are interpreted literally, because the object satisfies constraints by the verb in its base sense, are "he drink ink" and "ship disintegrate." Phrases not deemed literal (and not deemed metaphoric in a subsequent routine) include "ship plow sofa," "he drink sofa" and "ship sleep." The input "sofa drink wine," an anthropomorphism, is metaphorically comprehensible because the actor is not +ANIMATE, but is +CONTAIN, a feature requirement of anything that drinks. The interpretation for the metaphoric use retrieves the effect/result portion of the meaning of "drink" as discussed in 2.3, giving "wine start be in sofa."

Examples of test results for some PHYSICAL-domain metaphoric interpretations are given in Tab. 3.2. The examples do not yet include any connotations that make the metaphor interesting; rather they minimally extract the state or action being expressed.

Tab. 3.2: Test for physical-domain metaphoric interpretations

Phrase	Interpretation
sofa drink wine	wine START BE–IN sofa
ship plow sea	ship(FUNCTION:go) DO sea START (BE SHAPE:)
ship plow–through sea	ship GO THROUGH sea
ship plow sofa	none
sofa plow ship	none
he close wine	none

The phrase "ship plow sea" is interpreted, because "sea" satisfies the + 2-DIMENSIONAL constraint on something which plows, and "ship" is something with FUNCTION "go," while "sofa" is not. The same applies to "plow through," but here the focus is on the motion of the actor, not the change of state of the object; therefore the interpretation is different. The verb "close" is defined in the lexicon as requiring a +CONTAIN OBJECT (in the sense of being "hollow"), which "wine" does not satisfy; therefore "he close wine" is rejected.[11]

Results of tests for a few cross–modal metaphoric interpretations, to which some of the above comments apply, are given in Tab. 3.3.

Tab. 3.3: Test results for cross–modal metaphoric interpretations

Phrase	Interpretation
he close mind	he (IPART:mind) STOP POSSIBILITY–OF START think
prosperity disintegrate	[one] STOP have–wealth
country leap–to prosperity	country START have–wealth
prosperity leap–to country	country START have–wealth
he close prosperity	none
prosperity is occupied	none
prosperity leap–to sofa	none

11 An exception would have to be made for the verb "close," to allow *instruments* of closing as an *object* of closing, such as "door," "window," "lid," etc. as literal object.

This small set of examples is only "anecdotal," but suggests that a simple procedure with a small set of sense descriptors can result in interpretations of metaphoric phrases which are at least "on the right track," if not adequately detailed. On the other hand, one can expect that various kinds of uncertainties are bound to attend any computational attempt to distinguish a metaphoric from an incomprehensible phrase.

One kind is that of comprehensibility. For example, "prosperity leap to sofa" may be judged incoherent, but one can well imagine comments about a restored sofa that would support an interpretation. Another kind is that of unforeseen human linguistic usages and actions, though this may be addressable. To use the above example, given a reasonable size relationship, a FUNCTION of "go" for a verb is unnecessary for, say, a vase to "plow through" a window–if it is thrown. Still another kind of uncertainty is gradual changes of meaning, even of closed–set words, such as prepositions, which, as indicated in 2, are appended to verbs to form the targeted predicate. As Herskovits [11] notes, there are deviations from the "ideal" meanings assigned to prepositions. Meanings gradually shift to create slightly different or overlapping meanings, and comprehensible misuses result in tolerance of those uses. In addition, people may have slightly different meanings for words because of different associated experiences. Finally, ellipsis may affect comprehensibility judgments. The expression "he close wine" as it stands is incoherent. However, given ellipsis of the word "bottle (of)," it is comprehensible. Such complications are discussed in Chapter 5.

The foregoing illustrations of domains, structures and features to computationally interpret physical–domain verbal metaphor are of particular semantic interest when applied to cross–modal metaphor. Structures and domains, being orthogonal to one another, represent similarities and differences constituting analogies, and similarities between concepts in different conceptual domains invite further thoughts about what kinds of metaphoric extension can be made. Simple verbal components such as states, causation and conceptual domains usually do not distinguish verbs sufficiently to determine a choice for metaphoric extension. Actions are of different types, verbs have different connotations, and adverbial concepts sometimes play a role in the structure of a verbal concept, and are left for discussion in Chapter 5. As cross–modal metaphor involves non-PHYSICAL nouns lexicalizing concepts which are not "things," it is of interest to first consider how part–sentence metaphor may arise from the syntactic treatment of "abstract" concepts, as indicated in the following chapter.

References

[1] Arnheim, R.: *Visual Thinking*. University of California Press, Berkeley, CA (1969)

[2] Brooks, D.: The values question. url http://www.nytimes.com/2009/11/24/opinion/24brooks.html (2009). Accessed 24-11-2009

[3] Brugman, C., Lakoff, G.: Cognitive typology and lexical networks. In: S. Small, G. Cottrell, M. Tanenhaus (eds.) *Lexical Ambiguity Resolution,* pp. 477–507. Morgan Kaufmann, San Mateo, CA (1988)

[4] Clark, H.: Space, time, semantics and the child. In: T. E. Moore (ed) *Cognitive Development and the Acquisition of Language*, pp. 27–63. Academic Press, New York, NY (1973)

[5] Fass, D.: *Processing Metonymy and Metaphor*. Ablex, Greenwich, CT (1997)

[6] Fass, D., Wilks, Y.: Preference semantics, ill-formedness, and metaphor. *American Journal of Computational Linguistics* **9**, 178–187 (1983)

[7] Goodman, P.S.: Elizabeth warren is right: Jamie dimon needs to resign from ny fed. url http://www.huffingtonpost.com/peter-s-goodman/elizabeth-warren-jamie-dimon.html (2012). May 14, 2012, accessed 31-3-15

[8] Gruber, J.: *Studies in Lexical Relations*. Doctoral Dissertation, MIT, Cambridge, MA. Indiana University Linguistics Club, Bloomington, IN (1965)

[9] Hart, J.: B.c. Comic strip, *Field Newspaper Syndicate* (1978)

[10] Hayes, E., Bayer, S.: Metaphoric generalization through sort coercion. In: *Proceedings of the 29th Annual Meeting of the Association for Computational Linguistics (ACL-91)*, pp. 222–228. Berkeley, CA (1991)

[11] Herskovits, A.: Language and Spatial Cognition. Cambridge University Press, Cambridge, England (1987)

[12] Holt, J.: Self centered. *The New York Times Book Review* (2007)

[13] Honeck, R., Voegtle, K., Dorfmueller, M., Hoffman, R.: Proverbs, meaning, and group structure. In: R. Honeck and R. Hoffman (eds.) *Cognition and Figurative Language*, pp. 127–162. Lawrence Erlbaum Associates, Hillsdale, NJ (1980)

[14] Indurkhya, B.: *Metaphor and Cognition*. Kluwer, Dordrecht, The Netherlands (1992)

[15] Jackendoff, R.: Toward an explanatory semantic representation. *Linguistic Inquiry* **7** (1), pp. 89–150 (1976)

[16] Jackendoff, R.: *Semantics and Cognition*. MIT Press, Cambridge, MA (1983)

[17] Katz, J., Fodor, J.: The structure of a semantic theory. *Language* **39**, 170–210 (1963)

[18] Lakoff, G., Johnson, M.: *Metaphors We Live by*. Chicago University Press, Chicago (1980)

[19] Levin, S.: *The Semantics of Metaphor*. Johns Hopkins University Press, Baltimore, MD (1977)

[20] Ortony, A.: Beyond literal similarity. *Psychological Review* **86**, 161–180 (1979)

[21] Russell, S.W.: Computer understanding of metaphorically used verbs. *American Journal of Computational Linguistics* **Microfiche 44** (1976)

[22] Schank, R.: *Conceptual Information Processing*. North Holland, Amsterdam (1975)

[23] Schank, R., Abelson, R.: *Scripts, Plans, Goals and Understanding*. Lawrence Erlbaum, Hillsdale, NJ (1977)

[24] Semino, E., Hardie, A., Koller, V., Rayson, P.: A computer-assisted approach to the analysis of metaphor variation across genres. In: J. Barnden, M. Lee, J. Littlemore, R. Moon, G. Phillip, A. Wallington (eds.) *Corpus-Based Approaches to Figurative Language*, pp. 145–153. University of Birmingham School of Computer Science, Birmingham, UK (2005)

[25] Townsend, J.R.: *Written for Children*. Penguin Books, Ltd., Harmondsworth, Middlesex, England (1974). Citing L. Garfield, “The Strange Affair of Adelaide Harris,” Longman Group Limited, England (1971)

[26] Virgil: *The Aeneid*. Vintage Books, New York (1983). Translated by R. Fitzgerald

[27] Warne, K.: South african marine reserves. url http://ngm.nationalgeographic.com/2014/12/sa-marine-reserves/warne-text (2014). December 2014 issue, accessed 24-4-15

4 The Role of Abstraction

Verbal cross–modal metaphor in the syntactic/semantic format to be handled depends on abstraction in two different ways. First, the "object" is, conventionally speaking, "abstract." Second, in order to overcome the domain inconsistency between verb and object, the verbal concept must be generalized in the form of abstract components to accommodate its object (syntactic direct object or subject).

4.1 "Abstract" objects

Pullum [14], musing on a definition of a noun, recalls writing in a previous article,[1] "What we regard as a thing, or a kind of stuff, is determined by what our language provides us with nouns for." He cites a 1929 survey article on Vienna Circle thought [5] with the quote: "Ordinary language uses the same part of speech, the substantive [= noun], for things (apple) as well as for qualities (hardness), relations (friendship), and processes (sleep); therefore it misleads one into a thing–like conception of functional concepts." Pullum seconds this assertion by equating the imposed grammatical definition of a noun, via a lovely metaphor, as "trying to squeeze syntactic blood out of the turnip of naive metaphysics."

Nevertheless, the making of a nonobject into a grammatical object needs to be dealt with and plays a critical role in linguistic metaphor. Nonphysical nominals, especially those which do not have any spatial component, have not been widely treated in computational approaches to metaphor, except in specific domains. Such concepts are usually referred to as "abstract," e.g., "beauty" and "privilege." The word "abstract" more strictly refers to a property "abs–tract–ed from" (etymologically, "drawn away from"), a concept, such as the nominal "color," as opposed to, e.g., "performance" or "reason," which are rather "concrete." In this book, however, the term "abstract" is used in the looser sense, i.e., conventionally.

Words for nonphysical concepts are not as easily learned as physical concepts, which can be pointed out and then associated with a representation, such as a simple utterance to designate that concept. A word for a concept not physically representable could be invented through a word or phrase made up of spatial concepts called to mind by the concept. Within the culture such

1 "Being a Noun," Chronicle of Higher Education, June 2012

reminding would be shared by others, and the word becomes a dead metaphor. English is replete with words for nonspatial concepts that have a Latin or Greek etymology derived from spatial concepts. The metaphoricity of some of these words has been totally lost, as in the "speak" sense of the action "to express" (from Latin, "to press out"), or for that matter the word "metaphor" (from Greek, "to carry over/beyond"). The nominals generated from verbal or attributive concepts of this type are considered abstract.

It should not be surprising, then, that a very large class of metaphoric words, dead or "alive," is based on spatial concepts. At least some of these metaphorical words appear to be derived from the "embodiment" perspective as illustrated by Lakoff and Johnson [9] and researched by Gibbs [4], the premise that one's body, its orientation, etc. play a role in the metaphorical use of a word to describe some situation. Metaphoric *phrases* could be regarded as having a hint of this experiential aspect as well, though not usually spoken of as abstract or concrete, e.g., the idiomatic metaphoric phrase "to see light at the end of the tunnel" and the novel metaphor, "to sky-dive over the fiscal cliff."

For the type of metaphor under consideration, reification (or "nominalization")-treating an action, relation or attribute as an "abstract object" in the form of a noun-is a first step in the creation of cross-modal metaphor. Reification of an abstract concept allows a physical-domain verb to have an acted-upon object which is not physical. The idea that abstract concepts can be treated as if they were objects, such as suggestions, interest, efforts, etc., can be thought of as the "*abstract concept as object*" metaphor. [16] Reification enables us to understand, e.g., person A acting so that person B does not succeed, as A torpedoing B's efforts. In this example there is not necessarily any object that person A is targeting-unless of course what person B is trying to do is reified, that is, made into an "object," here "efforts."

The "abstract concept as object metaphor" also provides an explanatory basis for conventional metaphor themes, such as the "conceptual metaphors" of Lakoff and Johnson [9], a large class of conventional metaphoric language, much of which is dead or "partially frozen" and not noticed by language users/receivers. For example, Lakoff and Johnson's *MIND IS A CONTAINER* conceptual metaphor ("I can't get it out of my mind," "the thought was thrashing around in my mind") can be considered an instance of an (abstract) object being in a location (mind of human). Prior to its characterization as a *CONTAINER*, the mind is an "object" and, perhaps more significantly, what it "contains" is an abstract "object" as well.

One can reasonably imagine that the identified conceptual metaphors of Lakoff and Johnson and those in subsequent implementations by others do not cover all possible usages. The generality of the "abstract concept as object" metaphor facilitates metaphors not covered by recognized conceptual metaphors.

When these are encountered (see following chapter), they can be analyzed just on the basis of the particular words of the expression. In later work [8], Lakoff's comment that the "OBJECT" metaphor is instrumental in other metaphor themes amounts to consistency with this idea.

Reification thus provides a basis for more general spatial structuring, which can address unconventional as well as conventional metaphor. It does this by allowing physical verbal concepts to "operate on" abstract objects through abstraction, as follows.

4.2 Abstraction from verbal concepts

Cross–modal metaphor results when a physical verbal concept is used in an interdependent combination with an abstract object, i.e., a nominal in a different, nonphysical modality. Other domain differences are present in some metaphors, but physical–domain verbs are the most common, in accordance with reasons given in the previous section. An abstraction from the physical verbal concept which applies also to a verbal concept that could convey the expression literally is what endows an expression with a metaphoric character. For cross–domain metaphor, resolving the difference between domains of a verbal concept and its object represents a generalization of the physical similarity that makes within–domain metaphor possible, as explained in the preceding chapter.

In a production or interpretation of linguistic metaphor, *the abstracted representation of the verbal concept* can be thought of as an interlingua or common ground between its metaphoric and its (more) literal sense. For our purposes, the abstracted representation is transferred to the topic of the metaphor, dropping some details from its vehicle representation.

Some questions are, what does the abstraction consist of, at what level should its components be, and what form should their representation take? As a partial answer to the first question, as described in Chapter 2, Gentner [3] has demonstrated through psychology experiments that verbal *structures* are principally extended from vehicle to topic in metaphor; the abstracted underlying semantic structure is the most important component in a verbal metaphor analysis. But for many metaphors, adverbial concepts are also clearly present in the vehicle and are often the *point* of the metaphor. Expressions such as "He lurched (vs. glided) through the exam" and "They torpedoed (vs. chipped away at) his authority" certainly illustrate distinctions in interpretation. The above are all aspects of an expression calling for representation.

The question of level of representation components invites a look at the idea of a representation itself. In a sense, any representation, whether a mapping between real-world concepts and symbols or between those symbols and higher-level symbols, can be considered abstract. Kaput [6], in the context of mathematics/arithmetic learning, defines four interacting types of representation at various levels – 1) cognitive and perceptual, 2) explanatory representation involving models, 3) representation within mathematics and 4) external symbolic representation, such as a chip, which can be instantiated by various kinds of objects and can thus be a *generalization* or *abstraction* for cards, dollars, cookies, etc.

All of these modes to varying degrees are relevant to representations of the role of abstraction in metaphoric language, as well as other kinds of metaphor. The chips in type (4) can be thought of as an abstract representation corresponding either to real-world items in their natural-language categorization as "objects" or, at the next level, to words for those objects. The type (2) representations apply to models of science, language and metaphor. Given the analogy implied by Kaput's modes of abstraction, it might be proposed that the cognitive components which relate to mathematical abstraction, a much higher-level abstraction, are (or overlap with) those which structure linguistic metaphor. A consideration of whether math-like components might be suitable for an abstract representation system for metaphoric-language, then, suggests a consideration of the intersection of language and mathematics – itself a system based only on a consensus, but a consensus created by language users.

4.3 Mathematical language

Mathematics education research often refers to the power of mathematics to model many analogous situations through its abstract representation language. The two simple cases, "two scarves taken from five" and "two paintings taken from five" (or even "the loss of two *of* the five paintings") can both be represented mathematically by "5 – 2." This power of abstraction recognizably extends to metaphor as well. Consider the metaphoric usages in the following two examples. "There can be no silver bullet – in this case the top-down creation of a global carbon market – to bring about the desired end. But could there be a silver buckshot?" [7] "The stimulus is a 'silver buckshot,' not a 'single silver bullet"' [12].[2] Here both "silver bullet" and "silver buckshot" have the same general functions and similar abstracted features ("one" for a bullet, "many" for buckshot), used only for comparison. The wide applicability of "silver bullet," a

2 citing Vice-President Biden

frozen metaphor, is expected; "silver buckshot" is novel, yet is easily understood in both contexts. This is also a case in which the abstracted features (one, many, random) are math–like concepts. More significantly, however, these two different situations can both be covered by the metaphoric use of "silver buckshot."

To illustrate with the syntactically simpler format that is the object of our attention, the physical action underlying the verb construct "chase out/away" can be extended to apply to conceptually different types of objects. In the phrase "chase away three ideas," the "ideas," a reification of a mental concept, are "taken away" from the thinkers of the ideas; mathematically, to "chase away three mosquitoes," as in a word problem, may mean to subtract or "take away" 3. In both cases, a seemingly nonmathematical phrase ("chase away") is mapped to a more abstract phrase ("take away"), which can be mapped to mathematical subtraction. Symbols can be mapped from one domain to another through an abstraction representing, informally, "causing the absence of 3 items" (in two different domains).

Objects, abstract or not, that are abstracted to numbers, are of course not the only entities represented in simple mathematical form. Arithmetic equations can represent structures that include concepts that are not objects, such as frequency, as in "he fell *twice*." Equations also include *operators* that relate the sets; the abstraction to numbers of any nominalized thing allows it to be "operated on." This is similar to the case of linguistic metaphoric extensions from the physical domain, where verbal or attributive concepts may become abstract "objects," allowing other verbal concepts to "operate on" them in turn.

When we look at mathematical language more complex than "taken from" or even "chase away," we find that it is often difficult to characterize mathematical language and natural language independently in discourse. Mathematical concepts can be embedded in natural language, not only in mathematical word problems, but in ordinary language about situations.[3] For example, English embeds explicitly numerical references, such as "twelve" or "a dozen," as well as phrases that are mathematically relevant (even if one is not conscious of the fact in using them), such as "the rest of them," "another slice of," "altogether," "join" or "more than" [11]. These words all have simple (arithmetic) mathematical counterparts, rendering them candidates for primitive symbols for language representation as well. The meshing of natural and mathematical languages corresponds to Kaput's interaction between cognitive/perceptual and mathematical representations and suggests common components.

3 See Cummins et al. [2] for evidence of young children's confusion about such embedded language in word problems.

Prepositions, as a less obvious example, represent types of relationship that can be interpreted metaphorically as well as mathematically. Fig. 4.1 illustrates a way in which (a sense of) the preposition "of" can be understood linguistically as a metaphoric "part of" or quantitatively as "subtracted from." The extension of the preposition "of" in combination with syntactic reification or quantitative abstraction (middle level of ellipses in the figure) provides the basis for metaphoric and mathematical expressions respectively.[4]

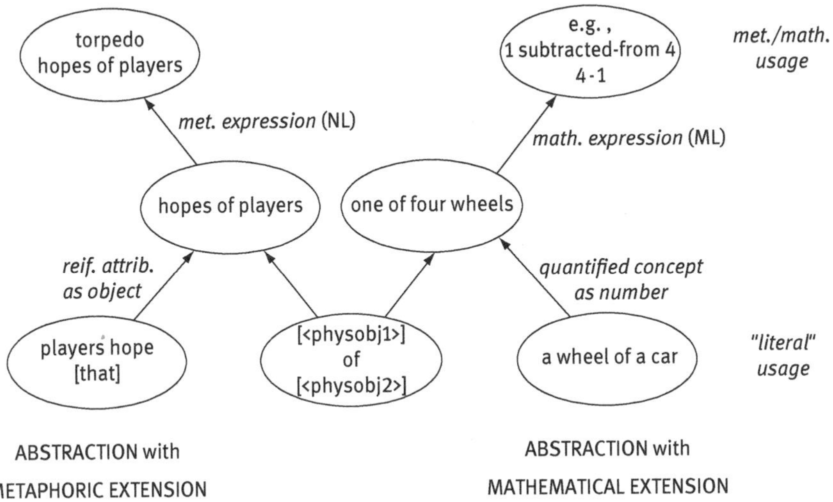

Fig. 4.1: Metaphoric and mathematical extensions of the preposition "of."

Reification, then, enables cross-modal linguistic extension in the way that quantification enables the mathematical extension. Much of metaphoric language can therefore be seen as sharing spatial grounding with mathematical language, suggesting that nonphysical as well as physical verbal concepts might be represented in terms of math-like components.

Abstraction between different levels provides an interesting connection between the assumptions underlying an ontology [17, 18] and the relationship of metaphor to mathematics in cognition. Lakoff and Nuñez [10] have described

---

4 See Brugman and Lakoff [1] for a systematic treatment of extended senses of many prepositions.

aspects of this connection comprehensively. While the main point here is not to make cognitive claims, the assertion of Lakoff and Nuñez that mathematics derives from the human cognitive system seems relevant to the illustrated relationships between abstraction, metaphor and mathematics. Their framing of this derivation does differ in their claim that mathematics *comes from* conceptual metaphors, while the present discussion (without any claims) characterizes mathematics and metaphor as parallel instances of metaphor. In both approaches, however, ordinary–especially spatial–ideas are seen as a grounding for metaphor and mathematics.

4.4 Representation

The answer to the second question asked earlier, as to what level of abstraction our representations should take, can therefore be guided in part by mathematical abstraction and its relation to language, mathematics offering the most "primitive" of possibilities. The caveat throughout is that representations of concepts are to some extent subject to the same inadequacies as word senses. As Panikkar [13] states, "The abyss between any word and its referent is more than ambiguity or ambivalence, but lies in the mysteriousness of the so–called referent itself." (p. 13). It should be kept in mind that "defining" even a particular word sense sets limits on it, which may be broken.

The complete abstract representation vocabulary in a specific form can be seen as an "abstract ontology," which brings up the question of the composition of such an ontology. The use of the term "ontology" in natural language processing differs somewhat from the classical definition; however, as Wilks [20] asserts, the distinction between the meaning of "ontology" in the classical and in the artificial–intelligence or NLP sense is not important for the purposes of AI or NLP. He also rejects any claims that "cleaning up" given ontologies will result in any notable advances in the field, a view which seems even more relevant to novel metaphor, with its potentially multiple meanings, than to NLP generally.

A cross–modal metaphor–relevant ontology is based necessarily not on any supposed objective reality, but on a certain view of the world through language, which itself differs between cultures and even individuals. An observation of Quine [15] concerning the ontology of language might be applied as well to abstractions *from* language to an abstract ontology–that is, that a difference between one person's ontology and another's are simply the result of differences in "slicing" or grouping of ontological parts. Thus one cannot expect that correspondences between ontologies will be one–to–one (cf. also Whorf [19]), no doubt an understatement as applied to NLP ontologies in practice.

The intent, then, in putting forward a particular ontology in the form of verb and nominal representations is not to justify the individual components, or to claim that the components of the abstract ontology are universal, complete, uniquely "correct," or language-free. It is rather to manifest the *role* of abstracted components in a computational metaphor interpretation system, together with suggested *types* of components. Relying on any such ontology may raise the "mentalism" criticism; however, as Quine said, "I know no better."

In line with these thoughts, the following chapters describe representations of abstracted components that support a rough computer interpretation of such phrases, specifically cross-modal metaphoric phrases.

References

[1] Brugman, C., Lakoff, G.: Cognitive typology and lexical networks. In: S. Small, G. Cottrell, M. Tanenhaus (eds.) *Lexical Ambiguity Resolution*, pp. 477–507. Morgan Kaufmann, San Mateo, CA (1988)

[2] Cummins, D., Kintsch, W., Reusser, K., Weimer, R.: The role of understanding in solving word problems. *Cognitive Psychology* **20**, 405–438 (1988)

[3] Gentner, D.: Studies of metaphor and complex analogies: A structure-mapping theory. In R. Hoffman (Chair), *Metaphor as Process*, Symposium conducted at the annual meeting of the American Psychological Association, Montreal (1980)

[4] Gibbs, R.: *Embodiment and Cognitive Science*. Cambridge University Press, Cambridge, UK (2006)

[5] Hahn, H., Carnap, R., Neurath, O.: The scientific conception of the world: The vienna circle. pamphlet (1929)

[6] Kaput, J.: Representation systems and mathematics. In: C. Janvier (ed.) *Problems of Representation in the Teaching and Learning of Mathematics*, pp. 19–26. Lawrence Erlbaum, Hillsdale, NJ (1989)

[7] Keim, B.: Post–kyoto: Silver buckshot, not silver bullets. url http://www.wired.com/wiredscience/2007/10/post-kyoto-silv/ (2007). Accessed 26-10-07

[8] Lakoff, G.: A figure of thought. *Metaphor and Symbolic Activity* **1**, 215–225 (1986)

[9] Lakoff, G., Johnson, M.: *Metaphors We Live by*. Chicago University Press, Chicago, IL (1980)

[10] Lakoff, G., Nunez, R.: *Where Does Mathematics Come from? How the Embodied Mind Brings Mathematics into Being*. Basic Books, New York, NY (2000)

[11] LeBlanc, M., Weber-Russell, S.: A computer model of the role of text integration in the solution of arithmetic word problems. *Cognitive Science* **20**, 357–408 (1996)

[12] Michel, A.: Biden touts the silver buckshot. url http://www.propublica.org/article/biden-touts-the-silver-buckshot-904 (2009). Accessed 4-9-09

[13] Panikkar, R.: *The Rhythm of Being*. Orbis Books, Maryknoll, NY (2010). Based on the Gifford Lectures, 1989

[14] Pullum, G.: A postcard from vienna. url http://chronicle.com/blogs/linguafranca/2014/12/02/a-postcard-from-vienna/ (2014). Accessed 2-12-14

[15] Quine, W.: *Ontological Relativity and Other Essays* (The John Dewey Essays in Philosophy). Columbia University Press, New York (1969)

[16] Russell, S.W.: Verbal concepts as abstract structures: The most basic conceptual metaphor? *Metaphor and Symbolic Activity* **4**, 55–60 (1989)

[17] Russell, S.W.: The role of an abstract ontology in the computational interpretation of creative cross–modal metaphor. In: B. Sharp, M. Zock (eds.) *Natural Language Processing and Cognitive Science (NLPCS 2008)*. INSTICC PRess, Portugal (2008)

[18] Russell, S.W.: Abstraction as a basis for the computational interpretation of creative cross–modal metaphor. *International Journal of Speech Technology* **11**, 125–134 (2009)

[19] Whorf, B.: Language and logic. In: J.B. Carroll (ed.) *Language, Thought, and Reality: Selected Papers of Benjamin Lee Whorf*, pp. 233–245. MIT Press, Cambridge, MA (1956)

[20] Wilks, Y.: Ontotherapy: Or, how to stop worrying about what there is. In: NLPCS, pp. 3–22 (2007)

5 Processing Cross–Modal Verbal Metaphor

This chapter describes how a program called MAP[1] (Metaphor Analysis Program), based on the earlier program of Chapter 3, supports the recognition and, more interestingly, the interpretation of cross–modal or cross–domain verbal metaphor in terms of minimal "more literal" paraphrases. Lack of detail or depth in paraphrases is offset by breadth of interpretation. As Hayes [11] argues, it is of little value to restrict one's research to apparent success within a toy domain; systems should aim for breadth, which in turn will make evident potential areas in which to develop more detail. The discussion of MAP to follow should reveal not only specific challenges to be addressed, but also some specific ways in which *any* paraphrase of a metaphoric example is often less than adequate.

In order to zero in on the semantic relationship involved in metaphoric extension, the wide variety of syntactic forms found in discourse, as in the previous program, is reduced to a simple syntactic input form. This simplification clarifies the semantic relation between the verb and its (abstract) direct object or its (abstract) subject. Articles/determiners are omitted and the verb is restricted to the infinitive form, as the concern here is only with the semantics of the open-set lexical items anchoring the metaphor analysis. However, possessive adjectives are allowed; they will be seen in Section 6.2 to occasionally aid in disambiguation. Specifically, MAP expects as input the syntactic form "subject verb [possessive-adjective] direct–object," or "[possessive-adjective] subject verb" In the former format the subject represents an animate being or an event acting on an abstract object, as in "eliza/news torpedo [his] hope." In the latter format the subject is the abstract object, as in "[his] hope crash."

Because of the heavy emphasis on semantic analysis in the interpretation of verbal (or any) metaphor, *representation* of the semantics of input verbs in the lexicon is the most important component of an implementation. The claim here is that in order to know what is extended from a metaphorically used verb to the topic, elements must be extracted from the verbal concept which reveal how people "likely" perceive the concept. Especially for cross–modal metaphor, where nonphysical concepts have no literal similarity to physical concepts, it seems reasonable that an ontology or set of components representing these elements should be such that a separation of perceived aspects of verbal concepts be preserved, and that the level of components is abstract enough to be seen as

1 MAP is programmed in Gnu Common LISP.

applying to both vehicle and topic. Thus abstract lexicon definitions are presented here before the interpretation process as a whole.

The remainder of this chapter first explains MAP's representation of differences and similarities between vehicle and topic in verbal metaphor in terms of nonextensible nominal domains and extensible verb components. Next its interpretation process is explained, including the role of its abstract lexicon and some means of distinguishing a coherent metaphor from an incoherent expression. Some problems presented by complexities of language are then reviewed. The chapter concludes with illustrations of output paraphrases.

5.1 Differences: Conceptual domains

As noted earlier, in metaphor there is a difference between the domain of the vehicle concept and that of the topic concept. The vehicle domain is the domain of the verb used in its putative literal sense. Some students of verbal metaphor point to the object nominal as the topic. It is arguably more proper to identify the verb in its extended sense as the topic concept, which of course is in the domain of the object nominal. MAP checks the relationship between the verb and its object or subject nominal for domain difference, which would mean that the input expression is not literally comprehensible and therefore *potentially* a metaphor, though this step is redundant, since the hosting NLU program would do this in triggering MAP.[2] In principle the recognition of verbal cross–modal metaphor is thus in concept the same as that of within-domain metaphor, with the difference that the PHYSICAL domain is only one of several conceptual domains, corresponding, as indicated earlier, to interdependent human faculties, such as thinking and perceiving.

In the lexicon, which consists of representations of vocabulary relevant to MAP's processing, each verbal and nominal concept is assigned a conceptual domain in which the concept is thought to be literal. Thus for "He plowed through Eliza's research proposal," the vehicle domain–the literal domain of "plow"–is found to be PHYSICAL; the topic domain as imposed by the object of "plow through," i.e., "research proposal," is mental–more specifically intellectual. "Plow through" is then considered to be of the same domain, i.e., mental, in a metaphoric sense. Of course, if the PHYSICAL sense of "research proposal"

---

2 The test for a literal expression before a metaphoric one is not meant to imply that metaphor is to be thought of as "deviant" with respect to literal language, or sought by humans only after literal interpretations fail.

is intended, and the multi-page proposal is strewn about on the floor, this example could be a within-domain, i.e. a PHYSICAL-domain metaphor, as both "he" and the paper "proposal" are PHYSICAL.

In determining the domain for which a nominal with more than one domain sense is literal, the question inevitably arises as to which usage of a nominal with different but metaphorically related senses is the literal one, i.e., the base usage from which metaphoric extensions are made. For example, the nominal "engine" might be considered either as basically PHYSICAL, since a common sense is that of a vehicle engine, or as more general, since it can be easily thought of as anything which causes something to function. The latter could be a metaphoric sense of the former, or the former could be a specific case of the latter. A bit of research could clarify the question, but is not needed, as the question is only of "academic" interest. The supposed metaphoric sense is quite frozen in the language and could of course be defined separately from the other sense. Of more relevance to the question of which should be considered the literal sense, an NLU program would be able to consider any one of a number of senses; it could then test to see whether the nominal makes sense in its defined domain. If so, it is understood; if not, MAP could test for a metaphoric sense, and it will also be understood. The domain of any sense that is a literal candidate, then, could serve as the base domain of the concept in the lexicon.

Recognition of a cross–modal metaphor through domain difference is in a sense easier than that of PHYSICAL-domain metaphor, since the differences between conceptual domains do not involve all the differentiating details of the PHYSICAL domain. However, working out and using a domain taxonomy to support recognition and interpretation takes some thought (and can never be claimed to be definitive), as the domains are interdependent as descriptors of human interaction with the world. The treatment of metaphor to be described utilizes a set of conceptual domains and subdomains for both nominals and verbs, with examples of nominals as follows. Some of these domains and subdomains are of more relevance to nominal metaphor, described in Chapter 6.

- MENTAL
  - INTELLECT ("idea," "data")
  - ATTITUDE ("anger," "celebration")
  - WILL ("ambition," "choice")

- SENSORY
  - SIGHT ("image," "view")
  - HEARING ("sound," "symphony")
  - TASTE ("flavor," "taste")

SMELL ("fragrance," "odor")
FEELING ("pain," "warmth")

- CONTROL (of)
  - EXTRINSIC
    - PHYSICAL concepts ("property," <any PHYSICAL concept>)
    - WEALTH ("bitcoin," "stocks")
    - ACTs deriving from authority ("rights," "duty")
  - INTRINSIC ("talent," "disability")

- PHYSICAL
  - ANIMATE ("Eliza," "bacteria," "Berlin," "ghost")
  - INANIMATE ("iceberg," "lens," "dirt")

Added for nominals, though not implemented:

- TIME ("week")

- SPACE ("attic")

Some examples of metaphoric extension between conceptual domains follow. Unless otherwise made explicit, "physical" refers to "physical–inanimate."

- Experts waded through questionnaires (physical to mental–intellect)
- The shows spewed forth zingers (physical to mental–intellect)
- The House guillotined the bill (physical to mental–intellect)
- The team sailed to victory (physical to control–intrinsic—but could be a pun on "sail")
- The characters floated through life (physical–inanimate to physical–animate)
- His reputation catapulted upward (physical to mental–attitude)
- The new building usurps their view (control–extrinsic [of act] to sensory–sight)
- Music flooded the room (physical to sensory–hearing)
- His ambition bloated (physical to mental–volition)
- Democracy slept (physical–animate to control–extrinsic [of act])
- They market democracy (control–extrinsic [of wealth] to control–extrinsic [of act])
- Chocolate agrees with her (mental–intellect to physical–animate)

(Note that extensions *to* animate beings ("chocolate agrees") can be considered to be anthropomorphisms.)

Of course, not all of these types of extension are equally prevalent. It is not surprising that most extensions appear to be made from the PHYSICAL domain–the most easily observed. (See also Lakoff and Johnson [14] for possible origins of metaphoric uses of a large class of physical-domain concepts.)

5.2 Similarities: Extensible verb components

The role of domain difference in the characterization of metaphor is relatively straightforward. It is the content extended between these domains, representing the existing or imposed similarities between vehicle and topic in a metaphor, which is of primary difficulty and interest. The abstract components representing this content are the concepts that, for purposes of MAP's implementation, language users are hypothesized in MAP to recognize in a literal meaning of a verb that allows them to understand a metaphoric use of that verb, even if they have never heard it before. In this way the MAP model differs from that of Lakoff and Johnson [14] as described in Chapter 2, which identifies conceptual metaphors already in common usage. As mentioned in the introduction to this chapter, the identification of potentially extensible components of verbs and their representation in the lexicon is the most important part of the interpretation approach.

The set of abstract extensible components, together with the conceptual domains, can be considered to be a cross–modal–metaphor oriented ontology, with the characteristic that the extensible components are orthogonal to the nonextensible domains. The structures and features as described in the following two sections either have a mathematical–physical counterpart as discussed in the preceding chapter and/or have a broad linguistic consensus.

5.2.1 Verb structures

The vehicle sense of a verb used in metaphor obviously differs from its topic domain sense. The question of what part of a verb's vehicle sense can be seen as part of the topic sense, i.e., what is *salient* to the extension, is one of the more challenging in metaphor analysis. As we shall see, salience is a particularly thorny problem for analysis of metaphorically used nominals, as discussed in Chapter 6. Verbs involve less uncertainty, in that a verb's underlying meaning *structure* must be maintained in an extension and is therefore always salient. Gentner and France [10], working in the area of psychology, note that verbs are mutable, that is, changed in accordance with the nominals with which they are used, resulting in polysemy (including metaphoric senses). However, this does not mean that a core part of the verb–its structure–is changed by its usage. In fact, Gentner [8, 9], in her experiments on verb recall, has found experimental support for the theory that verbs are stored as interrelated sets of components, suggesting that component (complex) relations, including affective relations, rather than (simple) attributes are extended in metaphor.

5.2.1.1 Objects and relations

In MAP, a STATE of an OBJECT is assumed to lie at the core of every abstract verb structure. It is the STATE portion of a verb representation which determines a domain specification; i.e., the domain of the STATE is the domain of the verb in the lexicon. In the absence of an AGENT as defined below, it represents what is *predicated* by an expression. Minimally, a nonagentive simple sentence with the given format can be thought of as consisting of a state which predicates either the existence, an attribute or action of an OBJECT, or two kinds of relationships of an OBJECT to another concept called a LOCATION. In addition, an attribute itself can have a value. These structures may take the following forms, with verb examples:

- (OBJECT BE) "exist"
- (OBJECT BE <attribute>)[3] "glow," "sulk," "hope"
- (OBJECT [BE] AT LOCATION) "occupy," "contain," "believe"
- (OBJECT [GO] THRU LOCATION) "traverse," "plow through," "ski down"

For the THRU case, which involves a temporal concept, the "LOCATION" is the interval (or multiple points) between and including static LOCATIONs in the sense of OBJECT AT LOCATION. For example, a person plowing through memories has the memories (the memories are AT him/her). LOCATIONs in nonPHYSICAL domains are generally animate beings.

In addition to the above STATE forms, MAP redundantly uses the form (OBJECT FUNCTION), which could alternatively be expressed as OBJECT BE, with the addition of a dynamic feature, or as OBJECT BE <attribute>, where the attribute is "dynamic," as for the verbs "live" and "be operative" (see 2.2).

These abstract STATE structures can be thought of logically as one- or two-argument predicates or as unary or binary abstract case structures. Here we can see some correspondence in the linguistic research of Bouchard [2], though Bouchard's purpose–to unite semantics with grammatical constraints–differs from that of MAP. The two case-like entities–OBJECTs and LOCATIONs, are similar to those of Bouchard. The addition of AGENT in MAP is unless otherwise indicated

3 Attribute values themselves, such as "red," "asleep" or "hopeful," supply specific properties of the world and are not considered abstract.

(a component of) an "external" cause, i.e., an AGENT not identical with a LOCATION of the STATE, arguably not a true case (see Section 2.1.2).

MAP's structures differ from traditional case structures, such as those of Fillmore, [7] in being semantic/conceptual rather than syntactic, and apply across all conceptual domains. They differ from both those of Fillmore and the "conceptual" cases of Schank [20] in that they are more atomistic. Fillmore's dative and locative cases, for example, and Schank's conceptual Recipient and Direction cases are combined in (abstract) LOCATION. In MAP, for example, a human recipient is (abstractly) a LOCATION of the received concept.

This elementary abstract formalization says that predicates can be seen as applying to OBJECTs described by themselves (in terms of being, functioning or having a quality) or described as attached to or possessed by animate beings, especially humans (in terms of being "at" the possessor). Thus for the verb "thrive," which has the basic STATE structure "OBJECT FUNCTION," a cross-modal metaphoric OBJECT (syntactic subject of "thrive") would be a nonPHYSICAL concept such as "memory." For an "OBJECT AT LOCATION" structure, the LOCATION is the animate "possessor" or "experiencer" of the OBJECT. For the verb phrase "gnaw at," as in "An idea gnawed at her," "idea" is the OBJECT AT the LOCATION "her."

In a relation, OBJECT and LOCATION are used to indicate the *role* configuration of the verb, i.e., whether the syntactic subject of a simple sentence serves as the OBJECT or the LOCATION of a STATE. In the example, "*Hope* inhabited his mind," "hope" is the OBJECT and "mind" as a part of "him" is the LOCATION. In the STATE structure of "*He* sheltered his hope," "he" is still the LOCATION, even though "he" is now the syntactic subject. A syntactic alternative is presented by input in which the OBJECT and LOCATION are reversed. For example, in the sentence "he wallows in his memories," "wallows in" appears to indicate that "memories" is the LOCATION and "he" the OBJECT. However, the fact that "he" is human and "memories" is a nonPHYSICAL concept triggers a switch of OBJECT and LOCATION, resulting in the structure "OBJECT (memories) AT LOCATION (he)."[4]

5.2.1.2 Further structural components

A STATE can be negated, started or ended:

- NOT <STATE> "sleep" (not "be awake")

4 Hobbs [12] noted this syntax-related metaphor, which expresses a variable as an entity at a location, in the form of the predicate calculus axiom "variable (x) & value (w,y,x) –> at (w,x,y)"; i.e., if w is the condition of y being the value of the variable x, then w is also the condition of x being at y.

- BEGIN <STATE> "arise"
- END <STATE> "collapse"

In addition to verbs predicating LOCative STATEs, there are verbs expressing an orientation or tendency towards or away from such a STATE:

- ORIENTED (OBJECT AT LOCATION) "head-for"
- ORIENTED (NOT (OBJECT AT LOCATION)) "lean-away-from"

An additional extensible component, HYPOTHETICAL, applies to verbs or verb forms that refer to a *possible* state, such as "be vulnerable" (possibly become hurt) and "be susceptible" (possibly become affected, especially by something bad or, ironically, good). The linguistic modal auxiliary "could" is thus implicitly represented by the component HYPOTHETICAL imposed on STATE structures, and may appear in their paraphrases.

Some quantitative attributes of OBJECTs, having an abstract nature, are also included in MAP, such as magnitude and comparison:

- MORE (OBJECT BE (MAG HIGH)) "grow" (intransitive sense)

An AGENT relates to the the causative component commonly included in linguistic and symbolic NLU systems. An AGENT *causes* a STATE. More accurately, a situation causes a STATE, where the situation may be a natural phenomenon or an action of an (animate) AGENT. Thus the AGENTive component of the structure of "cancel" could be the event "rain" or on the other hand the person/agent "organizer," where it is understood that the organizer *did* something to result in the STATE. In terms of the above HYPOTHETICAL feature, the verb "immunize" as a +HYPOTHETICAL verb can be analyzed informally as an AGENT doing something such that the immunized one *would not* receive a negative effect. A metaphoric example is "The tornado immunized her against fear of lesser disasters."

As mentioned above, an AGENT in MAP is external to the STATE structure unless otherwise indicated. For the verb "transplant," for example, the core structure is AGENT (OBJECT AT LOCATION), where a LOCATION may be a SOURCE or a GOAL. In the sentence "The judge transplanted the privilege to the opponents", "the judge" is the AGENT, "privilege" is the OBJECT and "opponents" is the GOAL LOCATION, the SOURCE LOCATION being implicit but unidentified. Here the AGENT is external, causing a change in the relationship between the OBJECT and the two LOCATIONs. This meaning of "AGENT" differs from that of AGENTs causing a transaction in which they themselves are a SOURCE or GOAL, such as "market (to someone)," where the actor is the SOURCE, or "swallow," where the actor is the GOAL. In "He swallowed the idea," for example, "he" is a voluntary recipient, i.e., an AGENT who is also a LOCATION, a GOAL LOCATION in this case. Thus AGENTs that transfer an OBJECT to or from *the AGENT's self* are distinguished in lexicon definitions from AGENTs that transfer an OBJECT between two LOCATIONs external to the AGENT.

Not all AGENTive verbs result in executed STATEs. An AGENTive verb may express an initiated action, which by itself implies uncertainty as to the outcome. For the metaphor "He marketed propaganda to the people," for example, the basic structure is:

– AGENT:he TRY (OBJECT:propaganda AT LOCATION:people)

Some verbs of course have more complex underlying structures and require nesting, e.g., to include the concept of purpose, or the use of closed-set connectives such as "and" or "then." The verbs "sift" and "sift through", for example, involve separation of two objects, where the purpose could focus on what sifts through or on what remains unsifted. For the example, "She sifted [through] her memories," the definition would have to include the concept of separation, but the point would seem to be her purpose. She is going to hold onto some memories and reject others. Such cases, involving more than one structure, have not been implemented.

5.2.2 Verb features

The underlying basic verb structure provides a skeleton for additional descriptors of any action represented by the verb, as well as others applying to the verbal concept as a whole. In addition to the nominal features presented in Chapter 3, *features* are used as potentially extensible descriptors of *verbs*. Being abstract, the features can cross domains, as structures do. With purely physical details left behind, the feature set is small.

Feature values are assigned to the character of the action, if any, represented by the verb; the effect on the actor or other semantic participant, if present; and the effect that might be evoked in a recipient (reader/hearer) of the metaphor. The following examples identify features, together with verbs exhibiting the corresponding feature values.

Action descriptors have a spatio-temporal character and are often easily imagined as pertaining to nonspatial concepts. These features have positive (+), negative (–) or variable (?) values.

- REPETITION (+/–) "shuttle"/"fly (to)"
- CONTINUITY (+/–) "glide"/"lurch"

The following feature is assigned only if the HIGH or LOW value is extreme and therefore salient:

- SPEED HIGH/LOW "race"/"crawl"

While the above features apply to the denotation of the verbal concept, features referring to *effects on an actor or observer (reader/hearer)* of the metaphor are connotative. The following can be assigned to both actors and observers:

- FORCE (HIGH/LOW) "plow through"/"glide through"
- EVALUATION (POSITIVE/NEGATIVE) "dance"/"lurk"

The features FORCE and EVALUATION invite more discussion, as they correspond to Osgood's [18] evaluation and potency factors – two of the three non-structural factors he empirically determined to be extended in metaphoric usage. (Osgood's third factor was "activity," which could conceivably play a role in MAP's ontology, being relevant to verbs such as "bubble" and "churn." In the current implementation, "activity" is partially accounted for by the descriptors SPEED and REPETITION.) See also corresponding experimental results on adjectives in work by Aarts and Calbert [1].[5]

FORCE, a physics concept in its literal sense, is used to express efforts of participants in a STATE or other intense effects on them. Thus FORCE-ON-ACTOR is an extensible component describing a common ground between the PHYSICALly used and the MENTALly used "plow through." However, as the OBJECT (after any role switch as described in the preceding section) is abstract, no extensible force on the *OBJECT* is posited. For example, in "the House guillotined the bill," the "bill" certainly does not experience anything. Instead, there is a FORCE-ON-RECIPIENT descriptor, which applies to a recipient of the *metaphor*, whether or not resulting from an forceful effect on an animate participant of the metaphor. Many examples designed to evoke force are given in a study of verbs used in sports headlines by

5 In the computational NLU field, these connotative responses can be linked with information and inference mechanisms used by Dyer [4], who has represented situations associated with affects in *nonmetaphoric* text comprehension. Three of Dyer's abstract structures representing emotional reactions are similar to those used in MAP, namely evaluation (POSITIVE, NEGATIVE), character experiencing the emotion and (scale of) intensity.

Smith and Montgomery [22], who considered whether there is any limit or pattern to the different ways sports writers can describe winning and losing. In their research, for example, Smith and Montgomery correlated an informally defined category of verbs, such as "pulverize," "crush," and "stun," with the case of winning by a large margin or in an upset. The verb thus is an attempt to represent reader feelings as well as or more so than the state of the team that is "pulverized," which may be very negative, but need be no different from that of a team that "lost." That is, while a team may experience defeat and reversed expectations, it is questionable whether it thinks of (or wants to think of) being pulverized, ravaged, demolished, guillotined, left in the dust, etc.[6] The verbs are rather chosen by sports writers as a result of their own (or of conventional) feelings or expectations and the goal of entertaining. However, the verb itself, with its literal connotations, conjures up the concept of FORCE (interpreted as "intensity" in the topic domain), suggesting an effect on the parties exposed to the metaphor.

EVALUATION is subject to considerations comparable to those of FORCE. EVALUATION applies to the structure as a whole, expressing a connotation that impresses the reader/hearer of the metaphor.

Another class of connotations extended in metaphor is provided by emotions, such as, say, the fear accompanying a torpedoing. Emotions are not abstract; they are real-world concepts which remain literal at topic domains, though in a diminished form as they affect the observer. (The fear experienced when a proposal is guillotined would not have the same intensity as that when a person is guillotined.) The role of emotions and perhaps EVALUATION is consistent with the pragmatics-related observation of Morgan [17] and others that metaphor "calls to mind" affective properties.

Finally, there is one verb feature, having to do with a voluntary disposition, which may be salient for human LOCATIONs, i.e., human sources or goals of a transfer. For a TRY verb, illustrated by "market" above, the success of the marketing, i.e., the attempted transfer, depends on the *willingness* of an animate LOCATION (goal/recipient) to accept the transfer. Thus the recipient of "market" is represented as *possibly* +WILLING (certain acceptance, e.g., implied by "sell," would obviate the need for the TRY component). For the verb "foist upon," the recipient LOCATION would be unwilling, i.e., –WILLING. For the (differently structured) verb "usurp," the source LOCATION of the usurped OBJECT, the one from whom something is usurped, is also –WILLING. This attitude, voluntary or not, however,

6 Some of these verbs are so close in their effects that it is dubious whether even human recipients distinguish their responses to such usages, especially when the verb has become assimilated in that context.

is not consistently extended for human actors. The actor who literally "harvests" does so voluntarily, but one who "harvests anger" presumably does not.

Most verbs will not exhibit all of these components. An example of a verb representation including some of these is given by the phrasal verb "plow through:"

Domain:
PHYSICAL
State Structure:
OBJECT THRU LOCATION
Role of Actor: OBJECT
Features:
ACT: (CONTINUOUS +) (SPEED LOW)
ACTOR: (FORCE HIGH)

5.3 Verb extension across domains

The extensible components spanning general spatial and nonspatial domains reveal analogies that enter into metaphor. Thus "learning" in the MENTAL domain has an abstract structure similar to that of "seeing" (in its instantaneous sense) in the SENSORY domain, "acquiring" or "earning" in the CONTROL domain, and "absorbing" in the PHYSICAL domain. The analogies hold for alternative syntax, such as the start of: an idea in one's mind, an image in one's sight, an object or money in one's possession, or a fluid in an object, respectively.

To make more visible the analogies between verbal concepts with similar structures in different conceptual domains, sample common verbs have been located in a matrix/grid (Tab. 5.1) according to their domains as rows and extensible structures as columns. Here just four columns representing four types of extensible structure are shown, with qualifying features ignored. With portions of the matrix in view, one might notice relationships between concepts underlying the verbs, see how various verbs are actually metaphoric assimilations now thought to be literal, and either think of verbs that might fit in blank slots, or else speculate why none does. As more than one verb can fill a slot (with some distinctions ignored), this orthogonally based schema implicitly provides a categorization of verbs.

A few clarifying comments about the individual verb entries and their relationships are in order, first concerning the senses of the entries. While some of the entries also have nonverb senses, a common verb sense is to be understood. Also, for AGENTive verbs, a verb in an OBJect slot is assumed to be used with a syntactic direct object; a verb in a LOCation slot, with a syntactic indirect object. Both objects may of course be present in a sentence, as in "She teaches them geography."

Second, examples in a given row are not always structural variations of the concept illustrating the STATE in the same row. The reason is mainly because of failure to find an appropriate candidate for a given column, which may be due to lexical gaps in the language or in the conceptual world. Thus some slots are blank or filled by alternative verbs illustrating the structure represented by the column. For example, while the nonAGENTive MENTAL–VOLITION verb "forget to" results in someone not doing something, the AGENTive "deter" in the same line may cause this result by another means.

Another way lexical gaps in an OBJ slot have been filled in the chart is with a passive version of a LOC actor verb or AGENT actor verb, e.g., "be lost" or "be stolen" for EXTRINSIC CONTROL of matter. The passive form simply represents the perspective of the OBJECT in a relation. Finally, lexical gaps may also be covered by assimilated metaphors, often with the conventional END-STATE CONTROL verb "lose." Of course, many verbs considered literal for nonPHYSICAL domains are still metaphoric from an etymological perspective, but the criterion is current usage, not origin.

A notation additional to LOC and OBJ, OBJVAL, illustrates attributive verbs with a positive VALue or connotation; for END-STATEs it indicates the loss of a positive VALue. Finally, I have used the psychology-based terms CP (Conscious Processor) and LTM (Long Term Memory) which Schank [20] uses to indicate active involvement and stored concepts respectively. An OBJECTIVE vs. SUBJECTIVE distinction for verbs such as "know" vs. "believe" has been omitted.

Tab. 5.1: Extension of verb structures across conceptual domains.

Domain	STATE	AGT BEGIN-STATE	END-STATE	AGT END-STATE
MENTAL				
–INTELLECT	LOC (CP) consider	LOC tell	LOC forget about	LOC distract
	(LTM) know	inform	forget that	disillusion
	OBJ	OBJ teach	OBJ become incredible	OBJ refute
	OBJVAL be true	OBJVAL verify	OBJVAL	OBJVAL
–ATTITUDE	LOC (CP) enjoy (LTM) like	LOC interest	LOC tire of	LOC bore2
	OBJ please	OBJ endear	OBJ start to bore1	OBJ
	OBJVAL be pleasant	OBJVAL enhance		

Tab. 5.1 (continued)

Domain	STATE	AGT BEGIN-STATE	END-STATE	AGT END-STATE
–VOLITION	LOC intend	LOC urge (to)	LOC forget to	LOC deter
SENSORY				
–SIGHT	LOC see	LOC show	LOC black out	LOC blind
	OBJ appear	OBJ show	OBJ disappear	OBJ blur
	OBJVAL be beautiful	OBJVAL decorate	OBJVAL fade	OBJVAL bleach
–HEARING	LOC listen to	LOC play for	LOC become deaf	LOC deafen
	OBJ sound	OBJ play	OBJ <sound> stop	OBJ mute
	OBJVAL be melodious	OBJVAL	OBJVAL	OBJVAL
–FEELING	LOC sense	LOC stimulate	LOC suffer	LOC torture
	OBJ feel (e.g., sharp)	OBJ sharpen	OBJ pain	OBJ inflict
–TASTE	LOC taste1 (try)	LOC regale	LOC	LOC
	OBJ taste2 (e.g., good)	OBJ	OBJ	OBJ
	OBJVAL	OBJVAL flavour	OBJVAL become stale	OBJVAL overcook
–SMELL	LOC sniff	LOC	LOC	LOC
	OBJ smell (e.g., good)	OBJ	OBJ	OBJ
	OBJVAL be fragrant	OBJVAL perfume	OBJVAL start to reek	OBJVAL pollute
CONTROL				
–EXTRINSIC				
OBJ = action	LOC be eligible	LOC enfranchise	LOC be pre-vented from	LOC disenfranchise
	OBJ be possible	OBJ permit	OBJ be prohibited	OBJ rescind
	OBJVAL be moral	OBJVAL		
OBJ = matter	LOC possess	LOC equip	LOC lose	LOC rob
	OBJ belong to	OBJ give	OBJ go missing	OBJ steal
	OBJVAL be of quality	OBJVAL		
OBJ = wealth	LOC own	LOC endow	LOC become bankrupt	LOC bankrupt
	OBJ belong to	OBJ donate	OBJ pay	OBJ confiscate
	OBJVAL be worth	OBJVAL revalue	OBJVAL depreciate	OBJVAL devalue

Tab. 5.1 (continued)

Domain	STATE	AGT BEGIN-STATE	END-STATE	AGT END-STATE
–INTRINSIC				
OBJ = action	LOC be able to	LOC train	LOC forget how	LOC maim
	OBJ be talent of	OBJ elicit	OBJ (ability) fail	OBJ frustrate
	OBJVAL be excellent	OBJVAL enrich	OBJVAL decline	OBJVAL
PHYSICAL –RELATIONS	LOC e.g., contain	LOC fill	LOC expel	LOC deflate
of identity	OBJ inhabit	OBJ insert	OBJ slip out of	OBJ squeeze out from
and proximity	OBJVAL be	OBJVAL emerge	OBJVAL disintegrate	OBJVAL incinerate
–ACTIONS	LOC function	LOC operate	LOC halt	LOC disable
	OBJ be <action of>	OBJ facilitate	OBJ become incoherent	OBJ disrupt

It can be seen that extensions between certain domains or subdomains range from totally assimilated to strange. On the other hand, these subdomain extensions could result in some creative metaphors that may not be widely recognized as coherent. Some extensions may produce figures of speech which only poets manipulating or even "violating" language for particular purposes might utilize.

5.4 Interpretation

The process of verbal metaphor interpretation in terms of a literal expression, as far as this is possible, depends partly on the type of verbal metaphor. Indurkhya [13] makes a distinction between metaphors based on a pre-existing similarity and those that *create* a similarity. Many metaphors are based on representations that capture similarities existing prior to use of the metaphor. By contrast, Indurkhya presents evidence of the significant role of similarity-*creating* metaphors in cognition.

For example, while the metaphor *The sky is crying* might be considered novel, it is not similarity-creating, since there is a recognizable similarity between crying (tears falling) and something that falls from the sky (rain drops); the *comparison* view would seem to apply to the metaphor, though the affect of sadness is additionally suggested by the metaphor. "The ship plowed the sea" also falls into this category.

Similarity-creating metaphor, on the other hand, is characterized as *change of representation*. As Indurkhya states,

> In instantiating the vehicle concept network in the target realm, parts of the realm are "grouped" together and made to correspond to the concepts of the concept network. In this process, the target realm is given a new ontology, and its structure, as seen from the more abstract concept network layer, is changed. (p. 254)

As an example of a similarity-creating metaphor, Indurkhya cites parts of a poem, "Seascapes," by Stephen Spender, that uses the image of a "downhill rush [of water]" to describe a swath of flowers on a hill. While perhaps the eye of the observer and the incline of the swath of flowers suggested the metaphoric image, there seems to be no inherent property in the swath of flowers similar to the movement of a liquid; the waterfall image is imposed.

MAP as well does not compare a vehicle representation with a topic, as its interpretation process does not assume an existing similarity. Rather, it is assumed that the abstract vehicle representation of the verb is imposed, i.e., directly projected onto the topic, altering its representation. If there is in fact an existing similarity, the imposition may simply be redundant. All metaphoric usages of verbs, not just similarity-creating or novel usages, are treated this way by MAP. One could argue that all cross–modal metaphor (which Indurkhya does not discuss) is similarity-creating, because the real-world details of vehicle and topic will always differ. In any case, as Indurkhya acknowledges, the difference between suggestive similarity-based metaphor and similarity-creating metaphor could be a matter of degree. The relationship might be seen as one of complementarity; the more the vehicle ontology is projected onto the topic, creating a similarity, the less the pre-existing topic representation remains.

The structure that is imposed onto the topic representation is given in MAP's lexicon. MAP's lexicon contains a small vocabulary (currently 90 verbs and 150 nouns, plus prepositions and possessive pronouns), but many verbs not included would have the same basic structure as some of the included verbs, differing in details and nuances. In the lexicon, a verb is assigned a conceptual domain and any subdomains, together with extensible properties i.e., its structure and other descriptors as described earlier. Nouns are also described in terms of their domains and subdomains, along with other properties relating to their metaphoric use, to be described in Chapter 6. All of these properties are formatted simply as attribute-value pairs.

Some sample definitions, with slight modifications for readability, are given as examples for several types of structure ("t" = "true"):

- CAUSE an OBJECT to BE <attribute>:
  (decorate (verb t) (domain sensory) (subdomain sight)
  (state (object be more-complex)) (change enter-state)
  (agent t)
  (evaluation positive))

- CAUSE an OBJECT to BE – AT ONESELF:
  (usurp (verb t) (domain co) (subdomain extrinsic act)
  (state (object at location)) (change enter-state)
  (agent to – self)
  (from unknown) (to self) (source -willing)
  (evaluation negative))

- BE – ORIENTED – TOWARD ENDing one's RELATION with a LOCATION:
  (lean-away-from (verb t) (domain physical)
  (state (object at location)) (change end-state)
  (direction t))

- BE in an INTERMEDIATE STATE between BEGIN and END:
  (plow-through (verb t) (domain physical)
  (state (object thru location)) (change nil)
  (continuous t)
  (force-on-actor high))

NOTE: "(change nil)" indicates a constant STATE (of "thru-ness"), not a lack of motion.

In interpretation of a metaphor, the extensible verb structures are transferred to the topic, subject to certain constraints discussed in the following section, and the domain of the verb is changed to that of the OBJECT nominal. This representation is still abstract, but can be mapped directly to English words, resulting in a rough paraphrase.

5.5 Metaphor vs. incoherence

Following verification of the input syntax, MAP checks the input expression for domain inconsistency between the verb and the abstract noun which serves as its object. Domain inconsistency indicates a possible metaphor. However, it is clear that not all phrases or sentences that do not make sense literally are comprehensible as metaphors. Metaphoric incoherence as determined by MAP may point to a misparse of an ambiguous sentence by the host NLU program.

Certain criteria, then, are needed to distinguish metaphoric expressions from those that are "anomalous" from the perspective of MAP, that is, not recognized as either possibly literal or metaphoric. Such criteria are of course not guaranteed to be reliable; even human judgment often varies widely as to whether an expression is comprehensible as a metaphor. One need only mention Chomsky's "anomalous" sentence, "Colorless green ideas sleep furiously," to have people jump in, insisting that if one thinks hard enough, the sentence has a perfectly good metaphoric interpretation.

Nevertheless, with respect to MAP's focus on the metaphoric coherence of verb and object, it is evident that different sentences with the same verb but different objects may elicit different judgments as to metaphoric comprehensibility. To "plow through a discussion or proposal" is easily comprehended, but to "plow through a word or a sight," not as easily, if at all. It would not seem reasonable to simply impose an abstract definition of "plow" onto an "interpretatation" incorporating either of the objects of the latter phrase. Thus a compatibility judgment between verb and nominal in MAP's subject–verb–object and subject–verb input, while imprecise, should be attempted.

More specifically, criteria are needed mainly to determine which verb and OBJECT combinations are metaphorically "coherent." Two observations regarding coherence can be made here. If a sentence is found to be metaphorically incoherent, it might be a coherent example of metonymy (see Fass [6] for a computational analysis) or some other linguistic trope. While the identification of other tropes are not pursued in MAP, the failure to find a literal or metaphoric relation can point to detection of some other trope. Second, as mentioned earlier, the sequence of words which a parser passes to MAP as a possible syntactic constituent of a sentence may not be such a constituent at all; the sequence may be incoherent, indicating a different parse.

One type of ambiguity subject to incorrect parse judgments is given by conjunctions, as in "Bystanders heard Erika break into the discussion and the subsequent decision of the participants." In this sentence "and" likely joins the objects of "heard" rather than potential objects of "break into," because a decision (in its sense as a simple action rather than as a written or spoken judicial decision) cannot be broken into, as it is not +CONTAIN.

The task for MAP's purpose is to determine those properties that the OBJECT must have for the imposition of the metaphorically extensible components of the verb to result in a coherent metaphor. One means is to define a set of descriptors of potential OBJECT nominals and the constraints imposed by verbal concepts on those nominals.[7] In most of the few feature-based theories of metaphor recognition, there is a set of descriptive features, some of which may be "violated" in passing from a literal to a coherent metaphoric interpretation and some not. Types of such violations and critical feature values, however, are not generally specified. Theories of metaphoric *in*coherence tend to be (even) more vague, though to some extent perhaps necessarily. Thomas [25], for instance, says, "When *too many* features are incompatible, a metaphor simply becomes preposterous" (italics added: the focus is on number rather than type).

MAP's descriptors can be referred to as "conceptual," in that they determine whether combinations of verbs and their object nominals are conceivable, whether conventional or not. While extensive experience will no doubt show the need for additional features, the set itself, with its lack of real-world detail, should remain small.

The application of the constraints determine whether the sentence is considered a coherent metaphor vs. a nonliteral, nonmetaphoric expression. One application of constraints is illustrated by Fass [5], who shows how "The ship plows (or plows through) the waves" can be interpreted metaphorically through a sense frame search which determines that both a ship and a plow "move through a medium." Explicit categories on the level of "medium," however, do not help in comprehending extensions to nonPHYSICAL domains, such as "The cashier was plowing through her memories," unless (nonextensible) physical details of a "medium" are dropped in its definition.

While Martin's [16] model differs from that of Fass, it appears that similar observations on the nature of constraints would apply. Martin's subcategorization of "process" in his *NONLIVING THING AS LIVING THING* metaphor-maps hierarchy makes sense, and a (nonliving) "process" may be metaphorically killed, as a living thing could literally be. However, he does not indicate what makes a process or, for example, lights metaphorically "killable," but not a floor or a broom. While one can say that the latter objects could be successfully subcategorized in a different place in the hierarchy, the general adoption of this strategy has the disadvantage of resulting in a complex system of overlapping categories. Rather, with an interest in the semantic knowledge of a concept, we may want to ask what characteristic or feature it is that makes lights seem lifelike and therefore "killable" and that might apply to other types/categories of object, like machines.

7 These are analogous to "selectional restrictions" as used in the field of linguistics for syntax.

5.5.1 Nominal descriptors

MAP's constraints on the OBJECT nominal are specified in terms of abstract features and two general categories. The categories correspond to the ACT and ATTRIBUTE "conceptual categories" assigned by Schank [20] to reifications of actions. Thus the noun sense of "leap" is assigned the conceptual category "ACT" and the noun sense of "color" is assigned the conceptual category "ATTRIBUTE."[8]

Features of abstract nominals are similar to those used for PHYSICAL nominals, but in being abstract comprise a smaller set. Physical features of nominals merge in nonPHYSICAL domains, because of the inconceivability or redundancy of distinct topological characteristics. Thus PART (of), FIXED/ATTACHED (to) and (be) CONTAINED (in) merge into a more general concept (representing, roughly, subordination) labeled by only one of these feature names–in MAP's case, CONTAINED. The abstract feature set is then left partly analogous (in a one-to-many relation) to the literal set. The 2-DIMENSIONAL feature disappears entirely in the abstract set, since an area-like quality does not mean much in nonPHYSICAL domains.

For cross–modal metaphor, such features, as they apply to concepts which are "objects" only in a metaphoric sense, are themselves metaphoric. As Tourangeau and Sternberg [26] note, features cannot be identical for concepts in different domains; at best they are analogous. The abstract +COMPLEX, then, can be seen analogous to the literal, physical +COMPLEX. For example, "plowing" an abstract concept implies thinking of that concept as +COMPLEX–in a *metaphoric* though assimilated sense. As a constraint, this feature value would be satisfied by a noun representing a composite concept (+COMPLEX), e.g., a "symphony," but not a simple one (–COMPLEX), e.g., a "(musical) note" or an ATTRIBUTE, e.g., "color." However, it is inevitable that there will be variation in human consensus regarding the application, or even the meaning, of such features.

Establishing such a feature set is guided by the criterion of "conceivability"–a more permissive criterion than that of what is normal or usual. That is, the constraints imposed by a verbal concept on its OBJECT in terms of these features indicate whether a phrase consisting of those two concepts is conceivable, absurd though it may seem. A minimal set of conceptual features in MAP relevant to cross–modal metaphoric coherence is:

8 The category FACULTY, which is an animate faculty/capability corresponding to a conceptual subdomain, such as "sight," also plays a small role in MAP, but does not contribute to theoretical underpinnings.

- +/−
- SHAPE (bounded, vs. amorphous, mass)
- 1-DIMENSIONAL (linear-like)
- FIXED/PART/CONTAINED (subordinate)
- COMPLEX (vs. elementary)
- CONTAIN (can contain)
- FLUID (all parts can move)
- ANIMATE (dynamic)
- REPRESENTATION (physical aspect of MENTAL–INTELLECTUAL or +ANIMATE concepts. See 5.5.2 and 5.6.2 respectively.)

A comparison of the following literal feature set with the above abstract feature set (with the exception of REPRESENTATION) may give an idea of how the abstract set is *abstracted* from physical properties.

- +/−
- SHAPE (vs. amorphous, mass)
- 1-DIMENSIONAL (linear–like)
- 2-DIMENSIONAL (area, surface–like)
- PART (of something)
- FIXED (attached to something)
- CONTAINED (2 types: surrounded by or embedded in)
- COMPLEX (vs. elementary)
- FLUID (liquid, gel or particulate vs. solid)
- ANIMATE

It should be kept in mind that these features reflect a naive characterization of objects in the commonly perceived world. In a scientific sense, for example, above the subatomic level all objects are +COMPLEX and no objects are +1-DIMENSIONAL; yet a spoon is perceived as −COMPLEX and a river as +1-DIMENSIONAL. Shifting our perspective to the other extreme, we know that the Earth moves around the sun; yet the sun is perceived as not only rising and setting, but also as being +CONTAINED in the sky.

Even between these extreme magnitudes, for PHYSICAL-domain metaphor, with its more complex relationships, perspectives involving size comparisons, i.e., between *actor* and object, play a role. One can easily conceive of “ants plowing through the post,” since to a small enough creature, any object of “plow through” may “look” +COMPLEX. Similarly, an object can “plow the sea” only if it can have a certain locative relationship with a sea and is at the same or smaller order of magnitude than the sea. Scale and granularity thus become relevant. A system to handle PHYSICAL-domain metaphor would have to allow for shifts in magnitudinal perspective, like "zooming" in or out. This capability would be similar in concept to Talmy's [23] identified cognitive process of "magnification" or "taking the close-up view."

While no claim is made for the psychological reality of these components, some of the features receive support through grammatical evidence. For example, the description level of these features is similar to that of the "cognitive notions" which Talmy [23, 24] noticed by observing grammatical specifications, mainly within the PHYSICAL domain. While Talmy is not concerned with semantic distinctions such as literal vs. metaphoric expressions or features, the above SHAPE feature, for one, directly corresponds to Talmy's "boundedness" feature. MAP's COMPLEX nominal feature, which is positive for concepts lexicalized as plural in number, can be linked to Talmy's "plexity" feature/category as applied to grammar. Talmy considers extensions of features of matter (PHYSICAL nominals) to actions, given that the appropriate grammatical elements are present. As an illustration of the "plexity" category, he designates as "uniplex":

MATTER: A bird flew in.
ACTION: He sighed (once).

He designates as "multiplex":

MATTER: Birds flew in.
ACTION: He kept sighing.

The verbal concept here is "multiplex," as indicated by the grammatical elements "kept" and "-ing," similar to the sense in which the nominal concept is "multiplex," corresponding to the element "s" pluralizing the nominal. The action "kept sighing" is not literally plural, but through extension from matter to action, we can think of this phrase as having a component in common with the concept of "plural." The feature "multiplex" therefore applies abstractly to the verbal as well as to the nominal form.

MAP's feature set, of course, differs from Talmy's, partly because MAP's focus is on semantic dependencies and constraints, while Talmy, generalizing his cognitive notions over nominals and verbs, emphasizes grammatically determined structural notions. The underlying theoretical concepts, however, are similar. If MAP accepted phrases such as "kept sighing," it would use REPETITION rather than the COMPLEX feature, but any reifying nominal of that phrase could have a +COMPLEX feature value.

The difficulty in establishing an absolute set of constraints may result in two problems:

- *Underconstraint.* Both the literal and abstract sets, being of small size, are liberal in allowing computational interpretation of novel metaphor prevalent in various kinds of discourse. A disadvantage of course is that such features may underconstrain interpretations.
- *Overconstraint.* Overconstraint may result, when unusual or creative but conceivable relations between verb and OBJECT are neglected in the determination of constraints. The interdependent problems of both overconstraint and underconstraint are further discussed in 5.6.1.

5.5.2 Constraints on coherence

Specifying constraints of a verb on its OBJECT is not usually as simple as listing a few feature values. Along with conceptual categories, conceptual domains/subdomains may also enter into constraint determination, since in spite of domain analogies providing the ground of a metaphor, there are also obvious nonanalogous distinctions. Of more complexity, constraints may include combinations or alternatives of all such descriptors and negations of either of these. There may be ANDs and ORs with mutual nesting. MAP's testing for verbal metaphor examines constraint definitions corresponding to all of these configurations as specified by the input verb. As a simple example of some of these, the verb "plow through" imposes on its OBJECT the following constraint (conditions for something which is propelled, e.g., a thrown vase plowing through a window, as discussed in Chapter 3, are omitted):

- AND
  - OR
    - +COMPLEX
    - +PLURAL
  - NOT conceptual category: ATTRIBUTE

In other words, a coherent metaphoric use of "plow through" would expect its object to satisfy the constraint specification: 1) be complex or plural and 2) not be an attribute. These conditions would exclude as an OBJECT the nominal "donation," for example, which does not satisfy the minimal constraints, as it is not +COMPLEX.

Given the necessarily vague nature of judgments of the applicability of such features to a nominal concept, as well as of the uncertainty in specifying constraints imposed by a verb in terms of those features, we might ask whether they are of any value in distinguishing metaphor from anomaly. It can be recalled that, as a case of alternative parse resolution, the +CONTAIN feature value is of use in the disambiguation of the earlier example, "Bystanders heard Erika break into the discussion and the subsequent decision of the participants." Since the phrasal verb "break into" imposes this feature value on the OBJECT and "decision" is −CONTAIN, "decision" would be selected as the OBJECT of the verb "heard" and not as the OBJECT of "break into".

Coherence judgments can be used to select word senses as well as alternate parses, as one sense may be "anomalous" relative to another. As one simple case of sense resolution, the +COMPLEX constraint by "plow through" imposed on its object accounts for the observation that "plowing through a letter" in the MENTAL sense of "letter," as in correspondence, is judged as coherent, while "plowing through a letter" of the alphabet is not. The incoherence of the alphabetic letter example would allow the acceptance of the correspondence sense of "letter" by the NLU host program.

Sense distinctions can be indicated not only by features but by domains. For example, the nominal "CD/compact disk" can be understood in two senses/conceptual domains, PHYSICAL and MENTAL−INTELLECTUAL; the PHYSICAL CD can be thought of as a MENTAL concept with a REPRESENTATION feature. "Skipping through the CD" makes sense metaphorically if the MENTAL sense of "CD" is meant, since it is +COMPLEX and the PHYSICAL sense (seen naively) is not.[9]

Another relevant disambiguation task in the context of computational text comprehension is that of determining the correct anaphoric reference, i.e., for pronouns such as "it," which may involve verb−object relationships between concepts that are separated in the text. Consider for example, "The discovery of the painting provoked an animated online discussion. Plowing through *it* was an arduous task." Here there is syntactic ambiguity as to whether "discovery," "painting" or "discussion" is the antecedent of "it." The nominal "discovery," as an instantaneous event, is not +COMPLEX as required by the verb "plow through," and is thus excluded. If we assume that "painting" has been identified as PHYSICAL by the host program, then "plowing through the painting" would be all within the PHYSICAL domain and the situation of the human discoverers "plowing through the painting" would be rejected as a PHYSICAL-domain metaphor. A view of a painting is (arguably) not +COMPLEX. "Discussion," on the

9 Or it can be considered as a "frozen metonymy" [6].

other hand, is +COMPLEX as required by "plow through," and is thus chosen as the antecedent of "it."

In terms of traditional linguistic theory, MAP's way of judging metaphoric coherence maintains the distinction between violated and preserved types of features, a characteristic missing from some feature-based systems addressing metaphor, such as that of Levin [15]. For example, Levin shows how the example "His courage evaporated" might be construed in terms of the automatic addition of an "abstract" feature into the selectional restrictions of "evaporate." He simply adds this feature to the allowable list when a "deviant" usage is encountered, with no judgment as to the comprehensibility of nonliteral usages including other abstract objects. This method holds even when the supposed deviance of a given expression, such as "hopeful sign," is not due to metaphoric extension, but rather represents a convenience-related transfer of "hopeful" by a speaker from the "hoper" to the object which instilled the hope. This approach stops short of allowing that expressions not recognized as literal may constitute some form of nonliteral language other than metaphor or some ungrammatical form.

A more notable characteristic of Levin's features, as well as those of the system of Drange [3], however, is their conflation of nonextensible features representing domains, such as +MENTAL, with extensible features. This conflation hides a theoretical grounding of metaphor analysis: MAP's representation of metaphoric extension through separation of extensible from nonextensible concepts allows not only a determination of literal incoherence but also an explanation of why an expression is a coherent metaphor.

5.6 Coherence problems

Language is a productive phenomenon, full of multiple meanings and imaginative expressions and continually subject to change. While some novel metaphors are paraphrased in their discursive context and therefore possibly interpreted with reliability by a program, those which pop up in isolation have no such support. Uncertainties and complications in the determination of coherent verbal metaphor interpretation must either be addressed or acknowledged as placing limits on computational language understanding.

5.6.1 Constraint fuzziness

The constraint problem, i.e., the possibility of underconstraint (resulting in wrongful identification of a metaphor) or overconstraint (resulting in missed interpretations) is related to the question of what is conceivable. One could argue that conceivability is inherently fuzzy and that drawing the line between conceivable and inconceivable dependencies is a futile endeavor. As Smith, Rips and Shoben [21] and many others have observed, category membership (and one could add, feature attribution and constraint) is often a matter of degree.

Drange [3], in considering the conceivability of expressions including PHYSICAL nominals, used literally, states that the "thinkability" or "conceptual meaningfulness" of a proposition is fuzzy only insofar as there is no agreement about the meaning of the constituent concepts themselves. Vagueness may come from unusual details, (e.g., an extension of a bed may have a bench attached, which would make "the seat of the bed" thinkable), or from a verb's crossing a literal vs. metaphoric threshold (e.g., it is literally thinkable that "squirrels, or even bacteria, like coffee," but only metaphorically thinkable that "plants like coffee"). Drange thus shifts the burden of determining the line of conceivability from the coherence of a proposition to the definition of individual concepts – basically the earlier discussed problem of the "fuzziness" of the application of features as descriptors of *abstract* concepts potentially used as OBJECTs.

Drange's argument seems to be true as far as it goes. In a cross–modal metaphor analysis, a "memory," for example, might be seen either as a simple unit or as an entire "scenario." As a unit, a memory is not easily conceived of as able to be metaphorically "plowed through," and can be excluded by the "+COMPLEX OR +PLURAL" constraint. But with "memory" as a scenario extending over time, the phrase "plow through a memory" (and in any case "plow through memories") could be accepted as coherent. This might be the case if "plow through [one's] memory" stands as such in the text and is followed by clarifying discursive context that refers to parts of "memory" as a scenario. Here the "thinkability" of the proposition does seem to depend on the interpretation of a constituent (OBJECT).

While Drange apparently believes that the thinkability question could be resolved if one could account for exceptions to a constituent nominal (such as "the seat of the bed"), the problem is (even) thornier for metaphoric expressions, especially cross–modal metaphor, since it is not only exceptions and different perspectives of abstract nominals that produce uncertainty. Verbs may be creatively extended in different ways, and may change with different objects. It is little wonder that Levin [15], considering this problem, is willing to simply allow everything.

Uncertainty of conceivability is only partially addressed in MAP by allowance for a "variable" value for features. However, as stated earlier, it should not be surprising that the determination of constraints and the strictness of their application may not be clear, since humans also often differ on whether a putative metaphor makes sense or not.

5.6.2 Complications

Inherent constraint uncertainty can be distinguished from a failure of constraints to account for unusual types of situations or for linguistic complications. One case of an unusual type of situation is that of exceptions to what is thought of as a norm and therefore not captured by the given features. While for PHYSICAL-domain metaphor, exceptions such as the "seat of the bed" are quasi-infinite and therefore illustrative as "inherent uncertainty," cross–modal metaphor is not subject to the quandary of dealing with physical details. However, there are some descriptors of nominals that, while generally applicable, do not always apply.

One type of exception concerns *representations* of objects. Through an anthropomorphic reference to a real-world representation on the basis of similarity of some components, originally metaphoric usages are often thought of as literal. Within the PHYSICAL domain, dolls can "open their eyes," which is easily accommodated. However, in its current implementation, MAP currently would judge as incoherent the verb–object phrase involving the creation of human faculties corresponding to certain subdomains, as in "(to) plant intelligence," because humans, at least, are already defined as intelligent.

This incoherence judgment would most likely be considered faulty by humans. If MENTAL–INTELLECTUAL capacities can be literally programmed into representations such as robots, then not only can robots be metaphorically thought of as thinking–or perhaps even "brooding," but one can "plant intelligence" into them. With input to MAP which included the object of "into" (e.g., "into R2D2"), MAP could easily account for such cases by placing constraints not only on the OBJECT, "intelligence," but also on the GOAL LOCATION, i.e., the electronic artifact. Thus while in MAP "(to) plant (or otherwise start) intelligence" would be excluded for animate beings, it would be accepted for robots, toys, images, etc., i.e., REPRESENTATIONs of ANIMATE beings. While other NLU programs could handle this exception in other ways, such cases remind us of how pragmatic considerations are not easily teased out theoretically from conceptual ones.

Complications other than constraint imperfections fall into at least three categories–potential multiple metaphoric interpretations; mixing of metaphor with other linguistic tropes; and changes induced by pronominal (possessive) adjectives. Multiple *literal* senses can be handled independently of one another. For example, the verb "lock up" taken literally can mean either to prevent someone from entering the locked up space or to prevent something from leaving it. Analogously, the metaphor "He locked up his mind," if based on the former literal sense, can mean to prevent anything from entering his mind – the conventional English meaning of closing one's mind; if based on the latter literal sense, to keep something in his mind from leaving. The former interpretation could be paraphrased as preventing learning, while the latter could be paraphrased as refraining from expressing something. The choice of interpretation would be left to the host NLU system with its access to the context of the expression.

The possibility of multiple metaphoric interpretations means that an initial failure to satisfy constraints for a coherent metaphor may indicate that a different metaphoric interpretation should be sought. Verbs may sometimes be interpreted in different though related metaphoric ways, depending on what types of constraints are satisfied by the OBJECT nominal. For example, the verb "swallow" with the object "idea" would be interpreted in terms of accepting the idea and thus holding the idea. The verb "swallow" with the object "pride," however, would result in an interpretation with the opposite meaning, that is, eliminating the pride and therefore *not* having the pride. In such a case the program can attempt disambiguation through the definition of "pride" as an attribute, i.e., something already belonging to the possessor, making acquisition redundant, while an "idea" may be new to the potential possessor.

A similar situation is presented by the verb "squeeze out," in that an attributive object, such as "pride," is generally perceived as internal to a person, thus being eliminated by being squeezed out, as pride is eliminated by being swallowed. A judgment of the metaphoric meaning of "squeeze out," however, may be less reliable than that of "swallow," partly because of the ambiguity of the preposition "out," i.e., does it mean "moved" or, alternatively, "extinguished"? The abstract structure of "squeeze out" in its literal sense simply means to move something out of something else, which is clear for *physical* concepts. The AGENT's intention behind its metaphoric use, however, may be either to rid someone or something of the object or to bring the object forth, thereby making it noticeable by other animate beings.

The current procedure of MAP is that if the object is an attribute (e.g., "He squeezed out her anger"), the interpretation is that the object is eliminated, while if the object is not an attribute (e.g., "He squeezed out the idea"), the result is that the object is drawn out. Cases such as this, however, indicate a more dominant role for the discursive context to resolve the ambiguity.

Other cases are equally or even more unclear, such as "squeeze-out a view", where the sense could either be as in "clouds squeezed out the view of the sun" or as in "he squeezed out a view between the two attendees in front of him". Articles/determiners may help to disambiguate such verbs. While "squeeze-out *the* view" may lead to the interpretation "to cause an existing view to end", "squeeze-out *a* view" more likely means "to cause the existence of a view." In such cases, MAP will not always get it right, but could offer more than one possible interpretation to the host program.

Alternative interpretations may also result, not from differing related phrase senses, but from the opaque quality of the syntax. For example, "anger" differs from many other emotions in that it not only is "possessed" by the subject; it is also directed at someone or something (see also Dyer [4]). A phrase conveying transfer of anger, such as "Eliza's anger made the rounds", may be interpreted either as adopted by persons other than Eliza or as directed toward various objects of the anger. The possibility that an emotion can be directed outwards is reflected in MAP's lexicon by a TARGET label associated with the emotion. As "anger" has a TARGET, the input sentence "His anger made the rounds" might mean that he aimed his anger at successive targets, rather than that successive people had his anger.

Another type of problem confronting semantically based verbal metaphor interpretation concerns the use of elliptical language, in which a word or words is omitted from a "more correct" syntactic expression. Nonproblematic examples are "if possible" for "if it is possible," or "The clothes were ironed, the dining room dusted." However, there is also a kind of "semantic ellipsis," which can affect coherence determinations by a metaphor analysis program when the ellipsis occurs between the verb and its object. Constraints must therefore be adjusted to allow for ellipsis. For example, the phrase "He sold the war" is incoherent if "war" is the OBJECT of "sell," but a coherent cross–modal metaphor if the elided OBJECT "idea (of war)" or "plan (for war)" is assumed as the true OBJECT, that is, if "He sold the (idea of, plan for) war" is meant.

Another example, which should *not* be interpreted as a metaphor, is "She heard a violin", where the more complete expression would be "She heard the sound of a violin [being played]", with domain consistency (SENSORY–HEARING) between the verb and its true OBJECT, i.e., "sound." Allowance for objects that can emit sounds render this expression literally coherent rather than metaphorically incoherent. These examples are all so conventional that it is not expected that metaphor analysis would be accessed; however, a program which depends on domain (in)consistency should "know" that apparent text incoherence does not necessarily imply either literal incomprehensibility or a need for metaphor processing.

Finally, The presence of a possessive pronominal adjective ("his/her/their") modifying the object can assist in disambiguation of phrases that potentially have differing though perhaps related interpretations. The interpretation of "break-into her discussion/anger," meaning "interrupt her discussion/anger," is different from that of "break-into discussion/anger," meaning "start to discuss/ be angry." The possessive adjective "her" in the former phrase indicates that the discussion already exists and therefore is not "started."

For the sentence, "he plowed through the research proposal", it is unknown whether he is (passively) reading or (actively) writing or orally delivering the proposal. In cases such as this, one of a few relational abstract (i.e., very general) verbal concepts is chosen, depending on the structure of the verb and the conceptual category and domain of the OBJECT. Here the output verb is INTERACT–WITH, which covers either interpretation. A possessive adjective referring to the object, however, as in "he plowed through her research proposal," allows an interpretation based on the passive meaning ATTEND–TO. Neither of these resolutions goes very far in providing a specific interpretation, but they do remove the metaphoric tension, offering the host NLU program general meanings which contextual factors can constrain.

Possessives do not always disambiguate verb structures, however. For the phrase "her defense," "defense" can be either the "defense" that "she" makes or the "defense" made on behalf of "her" (or both simultaneously). While MAP does not have the role of making this choice, it is able to produce either interpretation.

5.7 Paraphrases

MAP's paraphrases of cross–modal metaphor are intended to show its current ability to produce minimal interpretations consisting of "more literal" semantic content. Paraphrases are generated simply by a symbol-to-word(s) mapping from the abstract structure and features assigned by MAP to the input on the basis of lexicon definitions. For example, EVALUATION: POSITIVE is phrased as "with positive connotation." Sample paraphrases [19] are:

sentence: he plow-through elizas research-proposal
interpretation: HE CONTINUOUSLY WITH–EFFORT INTERACT–WITH ELIZAS RESEARCH–PROPOSAL

sentence: he torpedo elizas proposal
interpretation: HE SUDDENLY CAUSE NOT ELIZAS RESEARCH–PROPOSAL FUNCTION, WITH INTENSE EFFECT, WITH NEGATIVE CONNOTATION

sentence: eliza decorate idea
interpretation: ELIZA CAUSE START IDEA BE MORE–COMPLEX, WITH POSITIVE CONNOTATION

sentence: country prance-to prosperity
interpretation: COUNTRY EASILY START HAVE PROSPERITY, WITH POSITIVE CONNOTATION
sentence: his ambition blossom
interpretation: HIS AMBITION START BE GREATER, WITH POSITIVE CONNOTATION

sentence: his ambition bloat
interpretation: HIS AMBITION START BE GREATER, WITH NEGATIVE CONNOTATION

sentence: building corrode view
interpretation: BUILDING CAUSE NOT VIEW BE POSITIVE, WITH NEGATIVE CONNOTATION

sentence: democracy sleep
interpretation: DEMOCRACY NOT FUNCTION

sentence: he lean-toward anger
interpretation: HE BE–ORIENTED–TO HAVE ANGER

sentence: he market democracy
interpretation: HE TRY–TO CAUSE DEMOCRACY TO–GO TO WILLING UNKNOWN

sentence: he lean-away-from her defense
interpretation: HE BE–ORIENTED–TO NOT DEFEND HER

sentence: he squeeze-out her view (visual sense)
interpretation: HE WITH–EFFORT CAUSE SHE NOT HAVE HER VIEW

This kind of output is not detailed, nor can it be expected that metaphors (especially creative ones) can be adequately paraphrased. However, the fact that the paraphrases, while minimal, include the most essential content suggests that even if the ontology is significantly expanded, it will still be relatively small and transparent.

References

[1] Aarts, J., Calbert, J.: *Metaphor and Non-Metaphor: The Semantics of Adjective-Noun Combinations*. Max Niemayer, Tübingen, Germany (1979)

[2] Bouchard, D.: *The Semantics of Syntax: A Minimalist Approach to Grammar*. University of Chicago Press, Chicago, IL (1995)

[3] Drange, T.: *Type Crossings: Sentential Meaningless in the Border Area of Linguistics and Philosophy*. Mouton and Co., The Hague, Netherlands (1966)

[4] Dyer, M.: Affect processing for narratives. In: *Proceedings of the National Conference on Artificial Intelligence*, pp. 265–268 (1982)

[5] Fass, D.: *Met*: A Method for Discriminating Metonymy and Metaphor by Computer* (1989). Report CSS/LCCR TR 89-15, Simon Fraser University

[6] Fass, D.: *Processing Metonymy and Metaphor*. Ablex, Greenwich, CT (1997)

[7] Fillmore, C.: The case for case. In: E. Bach, R. Harms (eds.) *Universals in Linguistic Theory*, pp. 1–88. Holt, Rinehart and Winston, New York (1968)

[8] Gentner, D.: Studies of metaphor and complex analogies: A structure-mapping theory. In R. Hoffman (Chair) *Metaphor as Process*, Symposium conducted at the annual meeting of the American Psychological Association, Montreal (1980)

[9] Gentner, D.: Are scientific analogies metaphors? In: D. Miall (ed.) *Metaphor: Problems and Perspectives*. Harvester, Brighton, England (1982) pp. 106–132

[10] Gentner, D., France, I.: The verb mutability effect: Studies of the combinatorial semantics of nouns and verbs. In: S. Small, G. Cottrell, M. Tanenhaus (eds.) *Lexical Ambiguity Resolution*. Morgan Kaufmann, San Mateo, CA (1988) pp. 343–382

[11] Hayes, P.: The second naive physics manifesto. In: R. Brachman, H. Levesque (eds.) *Readings in Knowledge Representation*. Morgan Kaufmann, Los Altos, CA (1985) pp. 467–486

[12] Hobbs, J.: Metaphor interpretation as selective inferencing. In: *Proceedings of the 7th International Joint Conference on Artificial Intelligence*, Vol. 1, pp. 85–91 (1981)

[13] Indurkhya, B.: *Metaphor and Cognition*. Kluwer, Dordrecht, The Netherlands (1992)

[14] Lakoff, G., Johnson, M.: *Metaphors We Live by*. Chicago University Press, Chicago, IL (1980)

[15] Levin, S.: *The Semantics of Metaphor*. Johns Hopkins University Press, Baltimore, MD (1977)

[16] Martin, J.: *A Computational Model of Metaphor Interpretation*. Academic Press, New York (1990)

[17] Morgan, J.: Observations on the pragmatics of metaphor. In: A. Ortony (ed.) *Metaphor and Thought*. Cambridge University Press, Cambridge, England (1979) pp. 136–147

[18] Osgood, C.: The cognitive dynamics of synesthesia and metaphor. In: R. Honeck, R. Hoffman (eds.) *Cognition and Figurative Language*, pp. 203–238. Lawrence Erlbaum, Hillsdale, NJ (1980)

[19] Russell, S.W.: Map: An abstraction-based metaphor analysis program for overcoming cross-modal challenges. In: A. Neustein, J. Markowitz (eds.) *Where Humans Meet Machines: Innovative Solutions to Knotty Natural Language Problems*. Springer Verlag, Heidelberg/New York (2013) pp. 181–182

[20] Schank, R.: *Conceptual Information Processing*. North Holland, Amsterdam (1975)

[21] Smith, E., Rips, L., Shoben, E.: *Semantic Memory and Psychological Semantics*, vol. 8. Academic Press, New York (1974)

[22] Smith, M., Montgomery, M.: The semantics of winning and losing. Psychology report (1982)

[23] Talmy, L.: The relation of grammar to cognition–a synopsis. In: D. Waltz (ed.) *TINLAP-2 (Theoretical Issues in Natural Language Processing)*. Association for Computing Machinery, New York, NY (1978) pp. 165–205

[24] Talmy, L.: The relation of grammar to cognition. In: B. Rudzka-Ostyn (ed.) *Topics in Cognitive Linguistics*. John Benjamins, Amsterdam, Netherlands (1987)

[25] Thomas, O.: *Metaphor and Related Subjects*. Random House, New York, NY (1969)

[26] Tourangeau, R., Sternberg, R.: Understanding and appreciating metaphors. *Cognition* **11**, 203–244 (1982)

6 Nominal Metaphor

This chapter describes MAP's interpretation of nominal metaphor, i.e., expressions in which a noun is used metaphorically. While verbal metaphor may be an interesting way to describe a state or action, nominal metaphor is usually more complex, enlightening, entertaining – or insulting. Through the form "x is a y" or "x is the y of z," images, attitudes and bundles of putative facts are economically conveyed. Teasing out descriptions of these aspects is an important part of MAP's interpretation processing.

The chapter begins with a description of types and elements of nominal metaphor and the scope of MAP's processing. This is followed by a section on salience, i.e. what is being highlighted in a metaphor, and its representation for the above forms of metaphor. The general interpretation process for these types of metaphor is then presented, followed by some comments on addressing the coherence of nominal metaphor. The MAP program is then outlined, followed by sample paraphrases. An approach to metaphoric nominal compounds concludes the chapter.

6.1 The nature of nominal metaphor

The metaphoric use of a nominal occurs perhaps most often in the form "x <copula> y," where the copula or link posits the nominal x as the nominal y, as in "Juliet is the sun" (Shakespeare) or "the World – Wide – Web is a gold mine" There is an implicit analogy in these forms, which is made more explicit when a term is added, as in "billboards are warts on the landscape" [25]. In these metaphors, "x"is the target or topic and "y" is the source or vehicle. For the form "x is the y of z," "z" is in the domain of the topic. The analogy is completely explicit in four – term metaphors, as in Snoopy's metaphor, "Cats are the crabgrass on the lawn of life." Here the vehicle nominal receives elaboration, which itself spawns a secondary metaphor, "life is a lawn," with "life" as the topic concept. The four – term metaphor as reduced to three terms would be (the somewhat less comprehensible) "Cats are the crabgrass of life," where "life" is again in the topic domain. The analogy underlying the metaphor can also be expressed as "cats are to life as crabgrass is to the lawn." This form fleshes out the metaphor, but is of course more naturally expressed by Snoopy. MAP implements only the two – and three – term metaphors.

The copula "is/are" following the topic nominal bears special mention. since it suggests an equating of the vehicle and topic. Some philosophers of language, e.g., Searle [32], arguing from the perspective of pragmatics or the use of language rather than from semantics, have attempted to determine why a metaphor, which is not literally true, can mean what the originator intends it to mean. Morgan [24] points out that metaphoricity is not a property of a sentence, but rather what the speaker *does* in uttering the sentence. Consistent with this claim, we can theorize that a copula such as "is" in a metaphoric statement is not a declaration as in a literal subset relationship, but rather a command or *suggestion* to abstract certain properties from the vehicle concept according to one's experience. This assumption is realized in MAP through its imposition of abstracted vehicle–domain properties onto the topic concept.

Though the domains of the nominal senses linked in a nominal metaphor can be fairly easily identified, the characterization of a particular nominal metaphor as "cross–modal" is not as clear as in the case of verbal metaphor, where there is simply a difference in domain between the verb and object that appear in the phrase. The domains of the nominals do not tell us between which domains the extension is made. While some nominal metaphors are based on physical similarity ("the highway is a snake," "frying bacon is a snake," etc.) and can therefore said to be PHYSICAL–domain metaphors, other nominal metaphors, even when both vehicle and topic are physical, often rely on extension of properties in domains other than the domain of the vehicle nominal itself. In the metaphor, "the World–Wide–Web is a gold mine," it is not the physical properties of "gold mine" that are extended to the World–Wide–Web, but rather vehicle–nominal properties in other domains. A significant property of "gold mine" is (the acquisition of something of) high value, extended from the monetary (EXTRINSIC–CONTROL of WEALTH) value of gold to the informational (MENTAL–INTELLECTUAL) value of the Web.

It can be seen that for nominal metaphor as compared with verbal metaphor, an extra step is needed in the transfer process. Verbs have by their definitions certain structures and other semantic components that are consistently extended in metaphor. In a nominal metaphor, on the other hand, we do not know from the vehicle nominal at face value what is extended to the topic, since it has no obvious inherent structure. Verbal metaphor analysis is thus a necessary basis for nominal metaphor analysis, in that it is salient properties of the vehicle nominal–predicates that can be expressed as verbal or attributive concepts–that are extended to the topic nominal.

Because in MAP interpretations are based on imposition rather than comparison of properties, the same abstract property (something of high value) is transferred for the metaphor "dumps are gold mines" as for "the World–Wide–Web is a gold mine." For "dumps are gold mines," however, one can envision two different interpretations. What is obtained from a dump can be valuable PHYSICAL objects or valuable MENTAL–INTELLECTUAL objects. Thus there are two ways in which a nominal metaphor can be considered cross–modal: through domain difference of the nominals, as in "The Web is a gold mine," or through conceptual–domain difference of properties of same (conceptual)–domain nominals, as in "dumps are gold mines."

It must be kept in mind that the choice of properties to transfer to an interpretation is heuristic rather than certain. In contrast to a given verb sense, a nominal may mean many things to different people, and particularly to different cultures. One need think only of "water," with its many forms. Or of the many plants and animals that become metaphors for people: a wolf is a bad person in the U.S. but an independent person in Russia. There may even be a change over time: Within the U.S., the metaphoric meaning of "foxy" has changed. For the nominal metaphor, "dumps are gold mines," we may understand that dumps are posited as having something of value, but do not know without context whether it is the physical items in the dump or rather the information to be gained from them which are of value. In any case, if the experience that grounds the metaphor is not shared by the recipient, a novel metaphor will probably not work; a "lawn" as a vehicle nominal may fall into this category in, say, the Sahara. What is salient about a vehicle concept used in a nominal metaphor is, then, no trivial task.

6.2 Representing salient properties

As mentioned in Chapter 2, Ortony [25] describes the very character of metaphor as one of salience imbalance, i.e., features of high salience for the vehicle are of low salience for the topic. Thus for "the World–Wide–Web is a gold mine," the high salience of the value for "gold mine" is contrasted with the existing but lower salience of value for "World–Wide–Web,"[1] which then becomes recognized or strengthened through the metaphor. This characterization certainly seems to hold for many metaphors. Weiner's [40] computational interpretations of some metaphors according to Ortony's theory are focused on physical–domain metaphors where the salience imbalance is clear.

However, there are metaphors that work well, but where, as in "dumps are gold mines," there is no topic property of low salience to correspond with a highly salient property of the vehicle nominal. For this metaphor, in fact, the value feature of the topic, "dump," is negative.

Furthermore, as discussed in Chapter 5 with regard to verbal metaphor, there need not be similarity between vehicle and topic, as posited, e.g., by Miller [21]. The fact that both dumps and gold mines *contain* is perhaps a requirement

1 That the Web might be regarded as at least equally valuable simply makes it a relatively uninteresting metaphor.

for the metaphor to work, but cannot be said to convey the ground of the metaphor as a whole.

For many metaphors, then, we are left with the task of just determining what is salient about the vehicle nominal, but not necessarily the topic nominal. These characteristics can then be stored with a nominal concept as potentially salient, and are thus made available to the program mechanism for transfer to the output representation. As noted in Chapter 2, several computational researchers of metaphor have devised ways to establish the likelihood that a property is salient. Kilpatrick [19] stores putative salient properties with the vehicle nominal definition without worrying about low salience in the topic nominal, consistent with MAP's approach. However, he does not point to any guidelines in choosing such properties, which is a goal here.

In contrast to Kilpatrick, Carbonell [5] does attempt to identify, by ranking, which types of property, if present in the vehicle concept, are extended to the topic. His invariance hierarchy includes concepts that correspond to structural components (such as causation) used by MAP for verbal concepts. While useful, however, this enumeration stops short of a more general classification that points to why such properties make a nominal a candidate as a vehicle concept for nominal metaphor, i.e., what makes such concepts "remarkable" – what they call to mind.

Winston [43], though pursuing his theory with only a few examples, does take this step. His identification of importance, extreme value and distinction with respect to the class of an object relate to its remarkability. (Weiner [40] has integrated the latter two into her representations.) A concept may fit more than one of these criteria. For example, the height of the Empire State building in New York has extreme value; it is no longer the tallest building in the world, but its iconic status makes it distinct with respect to its class.

Winston's approach might be seen as complementing Carbonell's in serving as a guide for which properties are potentially salient. Significantly, Winston's criteria represent ways of having an *effect* on people and can be seen as subjective and perhaps stereotypical. To these criteria can be added *connotations*, similarly to the way they subjectively characterize verbs. Such criteria, then, do not necessarily depend on an objective definition of a concept, but rather what the concept means to people in their cultures, i.e., how the concepts are used or perceived.

Some of the properties which satisfy criteria such as Winston's appear in Carbonell's invariance hierarchy; the property which is perhaps the most important for artifacts is its function. The function of an artifact, as the reason for which the artifact was created, is the strongest candidate as a salient property [29] [30] (see also Rieger's [26] FUNCTION inference). FUNCTION appears in Carbonell's invariance hierarchy and satisfies salience criteria such as Winston's. Thus the function of a gold mine is more likely than e.g., any physical appearance to provide the ground of a metaphor in the absence of elaboration to the contrary. For natural objects such as glaciers, a "typical action" – what we notice about what the object does, e.g., advancing or receding, takes the place of function. Representations of persons can also include a "function" in the form of their occupations or roles in an activity, if they satisfy the above criteria, as in "He is the Henry Ford of organized religion," [15] where knowledge of Ford as the creator of the assembly line can be used to infer, minimally, that "he" does something resulting in the creation of a large church membership or attendance.

A lexicon editor defining a nominal must thus keep in mind how one typically relates to the nominal concept, if a program is to represent its metaphoric uses. This type of information could at least partially resolve difficulties seen (e.g., by Carbonell and Minton [6]) in the computer interpretation of "Sally is a block of ice." Here what is most salient is not that ice as an object is cold, but rather that one stays at a distance from ice (and therefore from Sally), with a negative connotation.

Such "second-order" descriptions, i.e. properties based on a relational implication or effect rather than on a simple attribute, can be challenging for an abstract – lexicon editor, but some subjective assessment of the semantics is necessary for analyzing novel metaphor. The focus on subjective properties and connotations specific to the vehicle concept requires a special mode of observation, which often includes thinking of concepts in terms of Gibson's "affordances" [12] – what does this object afford me, how would one stereotypically use, relate to or be affected by it?

Of course, stereotypes may also be thought of as prejudices enshrined in our language and culture. A didactic anecdote about a Zen master and his students, fanning themselves at tea, has him asking for the meaning of a fan. The first student gives a verbal definition as an instrument of cooling – our stereotyped definition – which is wrong. The second simply fans himself (better, presumably dispensing with the mere linguistic representation of fanning, but still wrong). The third places a cookie on the fan and passes it around – the correct answer. The desired stereotyped "answer" from the linguistic point of view is exactly what is sought to be counteracted by this lesson, illustrating the strength of stereotypical associations.

Nominal metaphor interpretations, even in a strong cultural context, are only likely, not definitive, since the speaker or writer may intend to highlight some less obvious aspect of the vehicle concept. Metaphors which have more obscure interpretations, however, typically require further elaboration, requiring multiple expressions, making it impossible for a lexicon editor to envision or foresee every possible experience with the object. As Johnson and Malgady [16] have suggested, it seems reasonable that one could judge how an individual *might* understand metaphorical usages of concepts for which one can identify highly salient characteristics in terms of attributes and affordances. In this case, we can assume that formally incorporating such judgments in a program will result in paraphrases of metaphor which *may* accurately refer to how the writer or speaker of the metaphor views the topic.

A potential further aid to salience identification is the immediate syntactic–semantic context of the vehicle nominal. For example, the most likely candidate for a salient property of a glacier might be the TYPICAL ACTION "advancing or receding," together with the associated attribute of the slowness and perhaps the inexorable nature of the motion. However, identifying this property as salient assumes that "glacier" is an *actor* in the expression; for a person "*on* the glacier" this property would be irrelevant. Direct contact with the glacier would select other properties, such as perhaps danger. These different types of situation can serve as a guide to salience candidates and can be distinguished within lexicon definitions. An informal metaphor-oriented definition of "glacier" is given in Tab. 6.1. Here properties which do not fit well under Winston's criteria are listed as "other;" i.e., they may be kept around for metaphors that are elaborated upon, but are not normally the first to come to mind.

Tab. 6.1: Partly formalized definition of the nominal "glacier." From "Information and experience in metaphor: A perspective from computer analysis," by Sylvia Weber Russell, *Metaphor and Symbol* 1(4), 1986, reprinted by permission of the publisher (Taylor & Francis Ltd).

Perspective	Salience Criteria	Properties
glacier is object		
	extreme value:	
	other:	large
		high
glacier is actor		
	distinct:	advance, recede (slowly)
	important:	
	other:	cannot be stopped
		change shape
		of land
		produce water
glacier is environment		
	important:	dangerous
	other:	cold
		still

This representation indicates that for an expression in which "glacier" is serving as an *actor*, its typical actions receive priority as a salient property. A more code-like representation with brief explanations of the "advancing and receding" predicate of a glacier is illustrated in Tab. 6.2.

Tab. 6.2: Abstract definition of "advancing–receding" context of "glacier."

Domain: PHYSICAL		
TYPICAL ACTION:		
Domain: PHYSICAL		
State Structure:		
	ALT STATE (OBJECT AT LOCATION) one-time alternating state	
	Role of concept: OBJECT (= glacier, moves between LOCATIONs)	
Features:		
	ACT:	
		(CONTINUOUS +)
		(SPEED LOW) (salient: extreme, distinctive)
		(FORCE HIGH) (salient: extreme)
		(REPETITION +) (ALT STATE repeats)

Tab. 6.3: Abstract definition of "gold mine."

Domain: PHYSICAL	
FUNCTION:	
Domain: EXTRINSIC–CONTROL (of gold)	
State Structure:	
	AGENT:GOAL
	BEGIN STATE (OBJECT AT LOCATION)
	Role of concept (gold mine): LOCATION:SOURCE
Features:	
	CONNOTATION: (EVALUATION (POSITIVE HIGH))

Whatever is metaphorically described as acting as a glacier, then, will receive an analysis corresponding to "going back and forth between states, continuously and slowly, with high force, where as annotated, not all properties are salient. For "glacier" as an environment, the subjective properties "cold" and "dangerous" are included, but not indicated as salient. (The danger would actually come from "crevasses" included in a more comprehensive definition of a glacier as a PART; the danger of this would then be transferred to the glacier as a whole.) An example of "glacier" as a vehicle concept used in discursive context is given in Section 6.3.2.

Salience principles, immediate linguistic context, and objective and subjective properties of the vehicle nominal itself, then, may all play a role in the choice of components to be included in a paraphrase for a particular usage of metaphor. Heuristics such as Carbonell's invariance hierarchy could have been of use in the present example; however, it is qualities such as distinctiveness which can supply a principled indication of salience and can be more successful. For example, the "natural tendency" (as Carbonell puts it) of a jungle, i.e., to grow, should not automatically take precedence over its "complex" feature, in spite of its relative position in the hierarchy, since many things grow.

The representation of "gold mine," as an artifact, consists mainly of its FUNCTION, as shown in Tab. 6.3.

A more compact format of this state structure is:

AGENT:GOAL
BEGIN STATE: (OBJECT:CONTROL–EXTRINSIC (of gold) AT LOCATION:AGENT)
EVALUATION (POSITIVE HIGH)

In contrast to an *external* AGENT, as defined in Chapter 5, the AGENT here is involved in the function transaction, i.e., serves as the GOAL LOCATION. As a resultant STATE is usually more important to a metaphoric interpretation than causative details, the means or instruments of obtaining such STATEs are not explicitly included, though they sometimes do play a role in metaphor.

6.3 Interpreting nominal metaphor

As expected, given the above observations, the paraphrase process for nominal metaphor is more complicated and open than that for verbs, mainly because of both lack of inherent structure of nominals and contingent choice of salient factors for transfer.

6.3.1 Example

MAP's interpretation of the input example, "World–Wide–Web is gold–mine," based on the representation for "gold mine" in the preceding section, is fairly straightforward. The abstract structure of the FUNCTION predicate and the EVALUATION are integrated into the FUNCTION predicate of "World–Wide–Web" (one takes information from it), giving the STATE structure and connotation in the topic domain as illustrated in Tab. 6.4.

MAP's output for "World–Wide–Web is gold-mine" is "World–Wide–Web is means of mental intellectual concept go from World–Wide–Web to user, has high positive connotation". The introductory phrase "means of" comes from the fact that "World–Wide–Web" is involved in a FUNCTION and can be thought of as an instrument. That the means is for a user is implied. The OBJECT is described only in terms of the domain, i.e., MENTAL–INTELLECTUAL, but could as an option provisionally be rendered as a generic intellectual object, such as "information."

Tab. 6.4: Abstract definition of "World–Wide–Web."

Domain: PHYSICAL	
FUNCTION:	
Domain: MENTAL–INTELLECTUAL	
State Structure:	
	AGENT:GOAL
	BEGIN STATE (OBJECT AT LOCATION)
	Role of concept (World–Wide–Web): LOCATION:SOURCE
Features:	
	CONNOTATION: (EVALUATION (POSITIVE HIGH))

The above paraphrase implies that the information comes from the Web, which in the immediate sense it does. However, with the Web as a means, one could also argue that the information comes from a (remote) originator of the Web pages, i.e., an AGENT who enables the user of the Web to obtain information. A comparison between the two different paraphrases (which would derive from two different but compatible representations) is a reminder that there may be more than one level at which concepts can be described metaphorically–or nonmetaphorically. In the former phrasing, "one takes information from the Web," the Web is a source location of the result; in the latter, "an agent transfers information to a user by means of the Web" or "one gets the information from an agent by using the Web," the Web is an instrument of communication from an agent. One can think of the Web, though an instrument, as a source at an immediate level and the originator of the information as a source at a remote level,[2] in which an AGENT applies an instrument.

2 This variability in perspective generalizes to many examples; we can think of some concepts either as sources or instruments (e.g., gold mine, the Web) and others as instruments or as part of a causing event in which an agent applies the instrument as an object to achieve a result (e.g., "key").

An argument can be made, however, that there are other artifacts, such as computer programs, for which creation by a programmer is quite irrelevant; programmers are not communicating with the user (at least not a human one), and to a lesser extent neither are Web creators. MAP therefore treats the current and similar examples without a "remote" source/creator.

6.3.2 Discursive context example

MAP does not currently process extended metaphors in discourse. However, a simplified analysis of a prosaic metaphor [39] involving "glacier" as a vehicle concept may suggest what kind of output is derivable from MAP's representations, how the determination of potentially salient properties may be further supported by words which refer to some of these properties, and what may be missing in the paraphrase:

> If political glaciers ever accelerate, this now seems the time: The slow movement toward an end of the Bonn coalition, still uncertain, still arrestable, has nonetheless become visible to everyone.
>
> The Christian Democratic victory and the heavy Social Democratic losses in state parliamentary elections in Lower Saxony removed the need for fine instruments of measure and left behind a rich terminal moraine of remarks and tactics. (p. 2)

The adjective–noun syntax of the "political glaciers" metaphor, while not currently handled by MAP, does not require different semantic processing. We therefore have a "political situations are glaciers" metaphor. The domain of the nominal "glacier" is PHYSICAL and that of the adjective "political" and the nominal "politics" is EXTRINSIC–CONTROL (of action).

"Accelerate," literally a PHYSICAL concept, represented as abstract components corresponding to "move with speed becoming greater," is a verbal predicate which applies to the TYPICAL ACTION of a glacier. The political situation therefore also has this representation. The situation up to this moment is described through the fragment "The slow movement...still uncertain, still arrestable." The "slow movement" is intended to merge the reader's images of the political situation and the glacier. "Toward the end of the Bonn coalition" is a literal segment continuing the narrative provided by the interpretation, with "toward" applying to both the vehicle, "glacier," and the topic, "politics." The "uncertain" and "still arrestable" qualifiers are also abstract enough to apply to both glaciers and to political and other situations. However, "arrestable" refers only indirectly to the glacier, as the opposite of "cannot be stopped" in the glacier definition. Reference to a negation or opposite of a defining element of a concept may indicate possibilities as to what is to come, reinforced by the word "nonetheless."

"Become visible" has no explicit counterpart in the glacier definition, as it is a step removed from the motion of the glacier/politics itself, but the discursive mode enables an extension of the metaphor. This phrase could also be viewed by itself as an extension from the SENSORY–SIGHT domain, metaphorically applicable to the MENTAL–INTELLECTUAL domain as recognition of the political situation and confirmation to the reader of the acceleration noted at the beginning.

With the rest of the piece the author delves into a rich elaboration of the glacier metaphor with further but nonsalient characteristics of glacial movement. Glacial moraines are familiar results of the movement of glaciers, with "remarks and tactics" explicitly indicated as topic concepts. The absence of a need for "fine instruments of measure" is an inference in the vehicle domain proceeding from acceleration.

Two observations regarding the informal analysis of this text come to mind. One concern is what is needed to computationally interpret the paragraph in its entirety. To interpret "removed the need for fine instruments of measure," for example, the knowledge base for the NLU program needs a more extensive description of glaciers, together with sophisticated inference–making and world–knowledge representation, some of which has been addressed by other areas of natural language processing. The second concern is that it would seem impossible to convey the full image of the political situation as a glacier by means of even a refined literal paraphrase of what is described here. A literal metaphor interpretation, it seems, can only represent the basic literal facts, not the experience of the reader who closely follows the descriptions of a situation.

6.4 Coherence vs. incoherence

As discussed in Chapter 5, a metaphor by definition is not (intended as) a literal expression. But obviously not every phrase which is not literally coherent is metaphoric. Recognition of the input as a *nominal* metaphor vs. an incoherent expression is not given much attention in MAP. This is in contrast to verbal metaphor, where coherence between the verb and object can be evaluated. Currently, therefore, a candidate for nominal metaphor that is input to MAP is assigned an interpretation, regardless of how comprehensible it might be judged to be.

However, three input cases are excluded from processing, i.e., from any determination of coherence or interpretation. If both nominals are +ANIMATE and there is no third nominal to further specify the assertion about the topic nominal, the expression is not handled; the input "Karl–Rove is Typhoid–Mary," (see paraphrase (29)) for example, needs further information (something is being spread, but what?). Second, partly to restrict its scope, MAP processes only nominal metaphors for which the vehicle nominal is PHYSICAL, as the majority of prosaic nominal metaphors appear to be based on a physical vehicle nominal. Potential vehicle nominals that are not physical are often questionably metaphoric, as in "play is pedagogy for children." Finally, if there are no salient predicates or features attached to the vehicle nominal, the input cannot be interpreted. Literal input in which one nominal is a subcategory of the other would presumably not be submitted to MAP, as the interpretation would be literal. Concepts in close, horizontally related subcategories, such as bacon and filet–mignon, can be metaphorically interpreted if the vehicle nominal has salient properties and the expression is a three-term metaphor, such as "fish is filet–mignon for cats."

The above exclusions from processing still leave a quasi-infinite number of nonliteral nominal combinations that would be accepted for processing, but could be incoherent as metaphors. If a predicate linked to the vehicle nominal is selected for transfer, the paraphrase resulting from imposition of salient properties has to make sense, at least to most people. Determining coherence is not an easy task, given human variability in understanding semantically complex phrases, but some considerations can be indicated.

The determination of coherence depends on whether the predicate(s) to be transferred from vehicle to topic can be comprehensibly integrated into the topic concept. Some (unimplemented) designs can be considered. If an abstracted salient structure of the vehicle nominal corresponds with a structure attached to the topic nominal, the phrase is probably metaphorically coherent (see the example of Fass and Wilks [9], "My car drinks gasoline," in Chapter 2). This is, however, a conservative way of judging coherence and is reminiscent of the comparison view of metaphor.

As an alternative, MAP could determine coherence only in terms of features. In the case of "knives are gold mines," both of the above approaches would find the expression to be incoherent. The structure–based approach fails, since the FUNCTION of a gold mine is for someone to obtain something, while that of knives is to be used to do something. MAP's feature–based criterion would fail, because knives do not CONTAIN.

If there is no corresponding structural predicate, but a correspondence of features does show that the abstract structure salient for the vehicle nominal is conceivable for the topic, then MAP's feature–based method, somewhat less reliably, can determine that the expression is metaphorically coherent. However, as mentioned with respect to the example, "Dumps are gold mines," in Section 6.1, there are often more than one coherent interpretations of a nominal metaphor based on different domains of a salient predicate. Dumps can be gold mines because of 1) what can be obtained physically from them or 2) because one can obtain information from them. The literal +CONTAIN feature value supports the first interpretation. In the second case, however, it is not a CONTAIN feature that enables the interpretation; the dump is a source, but only because the one who searches the dump does something (examine objects from the dump, think about them) to come up with the MENTAL concept, i.e., the information. In fact, many physical concepts could be the source of information, even if that is not their function. Such examples often have complex or natural (as opposed to artificial) concepts as topics, e.g., "geological formations are gold mines." The mechanism leading to the "information" interpretation is by itself, then, less direct and reliable than that of the "physical object" interpretation. In this case, the topic nominal is viewed as involving ellipsis of the phrase "study of [topic]."

There are also ellipses of other kinds of words, to be considered in the face of an expression that is metaphorically incoherent according to the feature-based criterion. "Knives are gold mines," for example, does not appear to be a comprehensible metaphor as it stands, because it is not +CONTAIN. "Knives" might really refer to "investment in knives." Thus if an expression with the format of nominal metaphor is found to be metaphorically incoherent, the possibility that the topic nominal is an elliptical usage as suggested by the discursive context could be considered.

The immediate lexical context, if available in the input, can give some clues to a more reliable interpretation. For example, the phrase "[some birds'] nests are gold mines for collectors" has a topic nominal, "nests," which has the +CONTAIN feature value, and "collect" has an abstract structure similar to that of "obtain." In this case a coherent interpretation is possible; if the definition of "nest," which as a physical concept could specify "eggs" and various animate creatures as the likely CONTAINed entities, the collection of eggs would be an easy inference, though it could be the nests themselves that are collected. If the immediate subsequent context "for collectors" is missing, the nests as "gold mines" could still be treated as a source of a (highly valuable) MENTAL-INTELLECTUAL concept, as for the "dumps" and "geological" examples.

While the examination of structures and features of topic concepts has not been fully implemented, MAP does produce rough but reasonable paraphrases. Its general mechanism can be sketched in the following section.

6.5 The MAP program for nominal metaphor

MAP accepts input having the syntactic form <nominal1> <verb> <nominal2>, where the verb currently is the copula (linking verb) "is," "is-a" or "are" or the simple possessive verb "has." Examples are "World-Wide-Web is gold-mine," "ideas are buoys" and "tax-code has weeds." Expected input also includes cases of three-term analogically expressed metaphor, i.e., where the vehicle nominal (nominal2) is followed by the third nominal in the domain of the topic, i.e., the domain of nominal1, connected by the prepositions "of," "for," "in," "on" or "over." An example is "[that veterinarian is the] Dr. Barnard of cows' horns" [4], input as "veterinarian is Dr-Barnard of cow-horns." MAP includes in its lexicon the salient (formerly widely known) property of Dr. Barnard in terms of transplanting human hearts. "Cow-horns" and other nominals with no evident salient characteristics are defined only with domain-and basic feature-information.

MAP then checks for various syntactic-semantic contexts in sequence (below), as given for the "glacier" example (Tab. 6.1 in Section 6.2), e.g., whether the vehicle is an object, actor, etc. The result is an accumulation of salient predicates (if there is more than one) for transfer to the topic nominal. The various types of context are prioritized for paraphrase heuristically (by introspection) for the vehicle nominal:

- as object
- as acted upon
- as instrument
- as actor
- as environment

The various case configurations are handled within procedures corresponding to each of these types. For example, the *actor* procedure interprets a case configuration including both 1) an external AGENT with an OBJECT, with or without SOURCE and/or GOAL and 2) a nonexternal AGENT with an OBJECT, with or without SOURCE and/or GOAL. Vehicle nominals as *object*s may have certain simple, potentially salient attributes attached to them: number, size, shape and homogeneity. For example, salad has a salient homogeneity attribute with a negative value, i.e., "mixed," applying to it as an *object*; Japanese beetles have a salient *object* attribute of (occurring in) high number, since they are usually – in some subcultures – *thought of* as being in swarms. In addition to serving as an *object*, they have a salient property applying to them as *actor*s, i.e., of causing a change in an object (plants). Confetti (which also has an *object* attribute of high number) has a salient property of being acted upon, namely, being transferred (i.e., thrown), in a random manner. A lens has a salient *instrument*al property of being used to obtain a particular view. Weeds have a salient *environment*al property consisting simply of a negative connotation. *Environment* refers to a concept external to the observer, which might be thought of as surroundings; thus if a vehicle nominal has a salient property as an *environment*, the paraphrase includes the topic nominal as "other/environment."

Within each procedure corresponding to one of these contexts, MAP accumulates the following parts of the abstract meaning of the phrase, expressed as English–like sentences. from the end of the paraphrase to the beginning. The first three parts are derived from the abstract definition of the vehicle nominal; the fourth part attaches the appropriate contextual wording:

- any given connotations
- case–related components
- predicative elements syntactically expected to precede the above components, together with any modifiers
- leading words such as the topic nominal and, if relevant, words specific to certain syntactic–semantic contexts, e.g., "means of" for instruments.

Case–related components depend on both the extensible state structure, possibly further specified as to SOURCE and GOAL in the case of a STATE that is OBJECT AT LOCATION, and the syntactic–semantic context(s) of the vehicle nominal. For example, salience due to the vehicle nominal as actor, e.g., "glacier" as moving in the "politics is glacier" example, would have an associated SOURCE and/or GOAL, though possibly UNKNOWN, but salience due to the vehicle nominal as environment would not.

Predicative elements may include causation, placement of a negation ("[is] not") of the verb, direction ("toward/away–from"), "try," and modifying features of actions, such as repetition, force on actor and continuity.

Possible leading words, following the topic nominal as subject are:

- a linking verb or variation of "have"
- "part of" for *objects* for which only a part is salient
- "means of" for *instruments*
- the phrase "concepts which" for *environments* possessed by the subject (see "tax–code" paraphrase (17) below)
- the phrase "other/environment" for *environments* without the preceding condition (see "politics is iceberg" paraphrase (5) below)

There is a small number of auxiliary functions having to do with domain idiosyncracies. For example, as there is a close association between visual and intellectual concepts in the real world, an instrument such as a lens, which has a SENSORY–SIGHT function, serves a MENTAL–INTELLECTUAL function, resulting in a transfer of a MENTAL–INTELLECTUAL concept to the topic. In addition, while MAP does not attend to minor grammatical elements in the paraphrases, certain "housekeeping" measures have been implemented, e.g., convenient use of the passive voice for certain cases, distinguishing input uses of the verb "to have" from forms of the verb "to be," and the adjustment of syntax for paraphrasing existential or other simple object attributes, such as number or size.

6.6 Paraphrase examples

Following are output paraphrases from MAP for a sample of input expressions, preceded by their original form and context, if derived from media. In conversion from original quote to input form, words have sometimes been replaced by nominals close in meaning to the phrase (see input (11) "words are confetti" and input (27) "East–West–dichotomy is lens for Ukraine") or have been omitted.

In this version of MAP, the symbol USER is used in some of the paraphrases instead of what in terms of structure would be LOCATION. It is used to refer to the person relating to the topic.

[Prompt for each input expression:] "Type a 3-word sentence (<noun1> is–a/is/are <noun2>) or a 5-word sentence, (add for/of/on/over/through <noun3>) in parentheses"

(1) sentence: (WORLD–WIDE–WEB IS GOLD–MINE)

(WORLD-WIDE-WEB IS MEANS OF MENTAL INTELLECTUAL CONCEPT GO FROM WORLD-WIDE-WEB TO USER, HAS HIGH POSITIVE CONNOTATION)

(2) sentence: (WASTE IS GOLD–MINE)

(WASTE IS MEANS OF PHYSICAL CONCEPT GO FROM WASTE TO USER, HAS HIGH POSITIVE CONNOTATION)

"This [document dump/U.S. State Department cables] is a potential gold mine for foreign–affairs scholarship" [8]

(3) sentence: (CABLE IS GOLD–MINE FOR SCHOLARSHIP)

(CABLE IS MEANS OF SCHOLARSHIP GO FROM CABLE TO USER, HAS HIGH POSITIVE CONNOTATION)

"If political glaciers ever accelerate, this now seems the time: The slow movement toward an end of the Bonn coalition, still uncertain, still arrestable, has nonetheless become visible to everyone." [39]

(4) sentence: (POLITICS IS GLACIER)

(POLITICS GO FROM UNKNOWN TO UNKNOWN WITH LOW SPEED, MAY NOT BE POSSIBLE TO STOP)

(5) sentence: (POLITICS IS ICEBERG)

(PART OF POLITICS IS OTHER/ENVIRONMENT WHICH IS NOT PERCEIVED, MAY BE DANGEROUS)

"Cultural pedagogy provides the lens through which Giroux and Pollock evaluate not only the media monopoly the Disney conglomerate has built, but also the impact of that media on the development of cultural attitudes and behavior through the targeting of youth,..." [33]

(6) sentence: (PEDAGOGY IS LENS FOR EVALUATION)

(PEDAGOGY IS MEANS OF PARTICULAR EVALUATION GO FROM UNKNOWN TO USER)

(7) sentence: (PEDAGOGY IS TOOL)

(PEDAGOGY IS MEANS OF MENTAL INTELLECTUAL CONCEPT BE)

(8) sentence: (PEDAGOGY IS GOLD–MINE)

(PEDAGOGY IS MEANS OF MENTAL INTELLECTUAL CONCEPT GO FROM PEDAGOGY TO USER, HAS HIGH POSITIVE CONNOTATION)

"Graduate education is the Detroit of higher learning" [37]

(9) sentence: (GRADUATE–EDUCATION IS DETROIT OF HIGHER–LEARNING)

(GRADUATE–EDUCATION IS LOW IN WEALTH, OPERATE NEGATIVELY)

"Detroit" could also be the conventional metonymy for the U.S. auto industry, which would of course be interpreted differently (or not, depending!).

"Graduate education is the Dubai of higher learning," [2] in response to the preceding:

(10) sentence: (GRADUATE–EDUCATION IS DUBAI OF HIGHER–LEARNING)
(GRADUATE–EDUCATION IS HIGH IN WEALTH)

"Faculty members born before 1980 grew up during a time when 'like' represented the beginning of a simile, rather than a piece of verbal confetti." [17]

(11) sentence: (WORDS ARE CONFETTI)

(WORDS ARE HIGH IN NUMBER, USED IN RANDOM MANNER, HAS POSITIVE CONNOTATION)

Because confetti is often associated with celebrations, it has received a positive evaluation in the lexicon, with unexpected, wrong results here (see discussion below).

"While men tend to applaud their spouses when they help to bring home the bacon, husbands aren't always as enthusiastic when women start bringing home the filet mignon." [38]

(12) sentence: (INCOME IS FILET–MIGNON)

(INCOME IS HIGH IN WEALTH, HAS HIGH POSITIVE CONNOTATION)

The negative reaction of the husbands to "filet mignon" is explicit in the quote, while "filet mignon" remains stereotypically positive for the women–and much of society.

"Periods [in text] appear like the Japanese beetles eating my knockout roses." [10]

(13) sentence: (PERIODS ARE JAPANESE–BEETLES)

(PERIODS ARE HIGH IN NUMBER, CAUSE NOT MENTAL INTELLECTUAL CONCEPT BE WHOLE, HAS NEGATIVE CONNOTATION)

NOT MENTAL INTELLECTUAL CONCEPT BE WHOLE would be clearer with inclusion of "text" in the paraphrase.

"Politicians left and right sprinkle it ['community'] through their remarks the way a bad Chinese restaurant uses MSG, to mask the lack of wholesome ingredients." [20]

(14) sentence: (WORDS ARE MSG)

(WORDS ARE MEANS OF MENTAL CONCEPT BE MORE–INTENSE, HAS NEGATIVE CONNOTATION)

"Jack Spicer saw language as a Trojan horse that could sneak in otherwise foreign ideas to audiences." [7]

(15) sentence: (LANGUAGE IS TROJAN–HORSE FOR IDEAS)

(LANGUAGE IS MEANS OF HIGH NUMBER OF IDEAS GO TO UNKNOWN, HAS NEGATIVE CONNOTATION)

"The Pythagorean theorem is an ancient oak in the landscape of thought. " [18]

(16) sentence: (PYTHAGOREAN–THEOREM IS ANCIENT–OAK OF THOUGHT)

(PYTHAGOREAN–THEOREM IS OLD, IS OTHER/ENVIRONMENT WITH POSITIVE CONNOTATION)

"One member [of a panel of tax experts] likened the current code to an untended garden. 'The problem is, the garden is growing nice flowers,' said another panelist, University of California, Berkeley professor Laura Tyson. 'So it's not a matter of weeds. These things were planted.' " [14]

(17) sentence: (TAX–CODE HAS WEEDS)

(TAX–CODE HAS CONCEPTS WHICH HAVE NEGATIVE CONNOTATION)

(18) sentence: (TAX–CODE HAS FLOWERS)

(TAX–CODE HAS CONCEPTS WHICH HAVE POSITIVE CONNOTATION)

"Neandertals were the SUVs of the hominid world" [44]

(19) sentence: (NEANDERTAL IS SUV)

(NEANDERTAL IS LARGE IN SIZE, WITH INTENSE EFFECT)

"tiny backbone living in corrosive swamp" to describe both a carp and any member of the House Ethics Committee [22]

(20) sentence: (CONGRESSPERSON IS TINY–BACKBONE)

(CONGRESSPERSON ACT WITH LITTLE EFFORT, WITH MILD/WEAK EFFECT)

(21) sentence: (HOUSE–COMMITTEE IS CORROSIVE–SWAMP)

(HOUSE–COMMITTEE IS OTHER/ENVIRONMENT WHICH MAY EVOKE FRUSTRATION WITH RESISTANCE, HAS NEGATIVE CONNOTATION)

(22) sentence: (SUN IS MAGICIAN OF LANDSCAPE)

(SUN CAUSE LANDSCAPE BE DIFFERENT, MAY EVOKE SURPRISE, HAS POSITIVE CONNOTATION)

In hurricane season, low–pressure systems over Africa are like eggs all in a line (adapted from comment by National Hurricane Center forecaster Gil Clark)

(23) sentence: (WEATHER-SYSTEMS ARE EGGS OVER LANDSCAPE)

(WEATHER-SYSTEMS ARE CLOSED IN SHAPE, MADE-NOT-WHOLE WITH LITTLE FORCE)

"Une operation miracle pour les...victimes de l'accident classique: cornes deboitées ou cassées. Grace au 'Dr Barnard de la corne,' elles pourront remonter sur le ring..." [4]

Translation: "A miraculous operation for victims of the classic accident: dislodged or broken horns. Thanks to the 'Dr. Barnard of horns,' they will be able to get back into the ring." (Dr. Barnard was a heart transplant pioneer.)

(24) sentence: (VETERINARIAN IS DR-BARNARD OF COW-HORNS)

(VETERINARIAN CAUSE COW-HORNS GO FROM ANIMAL TO ANIMAL, HAS POSITIVE CONNOTATION)

"If Boston is the Athens of America, then the Silicon Valley, northern California's microelectronics mecca, could be considered the Florence of the Information Age" [27]

Here the supposed conditional merely sets up the conclusion, rather than supporting its interpretation, as the general categories of the third terms ("America" and "Information Age") do not correspond in category. However, the perception that Florence is/was outstanding or important in some way grounds its salience.

(25) sentence: (SILICON-VALLEY IS FLORENCE OF INFORMATION-AGE)

(SILICON-VALLEY CAUSE INFORMATION-AGE CONCEPT BE NUMEROUS, HAS POSITIVE CONNOTATION, IS OTHER/ENVIRONMENT WHICH IS PERCEIVED WITH HIGH POSITIVE CONNOTATION)

"In his past writings, Cage has delighted in putting his most sober ideas and his most whimsical notions together in a salad bowl and tossing them wildly." [13]

"Toss wildly," a verbal metaphor, provides entertainment and a nuance that would be difficult to capture in literal paraphrase.

(26) sentence: (IDEAS ARE SALAD)

(IDEAS ARE MIXED IN HOMOGENEITY)

"The media look at Ukraine through an East-West lens"

(27) sentence: (EAST–WEST–DICHOTOMY IS LENS FOR UKRAINE)

(EAST–WEST–DICHOTOMY IS MEANS OF PARTICULAR CONCEPT OF UKRAINE GO FROM UNKNOWN TO USER)

"Big Data as a Lens on Human Culture" [3]

(28) sentence: (BIG–DATA IS LENS ON CULTURE)

(BIG–DATA IS MEANS OF PARTICULAR CONCEPT OF CULTURE GO FROM UNKNOWN TO USER)

"Karl Rove is the Typhoid Mary behind the plague of dark money" [41]

(29) sentence: (KARL–ROVE IS TYPHOID–MARY OF MONEY)

(KARL–ROVE CAUSE MONEY GO FROM KARL–ROVE TO UNKNOWN, HAS HIGH NEGATIVE CONNOTATION)

A fuller definition of the salience of "Typhoid–Mary" would easily produce a better paraphrase in terms of going to *many* unknowns. And the more complex addition of elements representing passage from one unknown to another, and from there to another... would be even more consistent with the vehicle, "Typhoid Mary."

"It is the East, and Juliet is the sun" (Shakespeare, *Romeo and Juliet*)

(30) sentence: (JULIET IS SUN)

(JULIET IS OTHER/ENVIRONMENT WHICH MAY EVOKE WARMTH–AND–JOY, HAS POSITIVE CONNOTATION)

"ideas which seem at first glance to be obvious and simple, and which ought therefore to be universally credible once they have been articulated, are sometimes buoys marking out stormy channels in deep intellectual seas." [42]

(31) sentence: (IDEAS ARE BUOYS)

(IDEAS ARE MEANS OF MENTAL INTELLECTUAL CONCEPT GO FROM IDEAS TO USER)

This "paraphrase" is obviously only partial, because this version of MAP does not consider the specifically intended sense of "buoy" or the remainder of the author's metaphor (see improvement suggestions below).

sentence: QUIT
GOODBYE

It can be seen from these examples and notes that the successful (and some less successful) aspects of the paraphrases derive from simple representations of vehicle concepts. The "sun is magician of landscape" metaphor is a good example of what a metaphor can do better than a similarly short but literal paraphrase. Nearly everyone has noticed the different look of a landscape in early morning darkness from that of its evolution into light. Surprise may no longer enter into the experience. But magicians usually do evoke surprise, and the image of the sun as a magician can "reignite" the wonder of visual transformation.

Obviously there is room for improvement in most of these paraphrases. There are several ways in which inadequacies could be addressed. Most basically, from the perspective of knowledge representation, one could revise salience judgments for the vehicle nominal in the lexicon and/or distinguish related from unrelated (or only contingently related) properties. For example, a tool is currently defined as a means of creation, but could of course additionally be defined in its related sense as a means of repair (causing a NOT WHOLE object to be WHOLE). Both properties could be realized as related alternatives in the paraphrase and would therefore probably not be wrong, though a specific intention in the discursive context might be missed.

Different *types* of properties, however, such as those representing physical form vs. use by or effect on people, may or may not apply together in an interpretation. The "words are confetti" example is a case in which three salient properties are represented in the paraphrase, but one of them – the positive connotation associated with its typical celebratory use – is questionable. While the celebration aspect of confetti depends on its physical manifestation of being high in number (and small, though this property is not represented), the relationship does not apply conversely. The insertion of a modal indicating only possibility (such as the word "may") appears to be necessary when connotations or relationships with humans are salience candidates for a nominal concept. This type of nuance does appear in many of the other above paraphrases. It is interesting, however, that although the author of this quote is not celebrating the use of the metaphoric confetti, she may be suggesting that the "thrower" of the verbal confetti is doing so with glee, thereby raising the entertainment value of the metaphor.

The rejection of an inconsistently salient property for an initial interpretation of a metaphor, however, does not mean that such properties should be ignored in extended discourse. People discussing something often counter the metaphoric point made by a conversation partner by referring to a potentially salient property of the vehicle concept unrelated to the given one, sometimes as a joke. Thus Mark

Shields[3] facetiously replied to his counterpart's characterization of Pres. Bill Clinton as Rasputin (he comes back to life again and again) by noting the characterization of Rasputin as "diabolical" – a property hardly related to "coming back."

In a similar dialogue referring to alternative properties, a metaphoric usage is cleverly paired with a literal *nonsalient* property. A Dilbert comic strip [1] has the boss referring to a "core" of an apple metaphorically as the literal central positive values that the firm is to respect, while Dilbert notes that the literal core is what is literally and – it is suggested – metaphorically thrown away. As Dilbert's reply deviates from the frozen association of "core" with "center" in favor of an incidental literal property, it is a surprise – and makes the point.

Finally, it is clear that an analysis of the discursive context plays an important role in the interpretation of many metaphors and is often necessary to arrive at a disambiguated interpretation. The original context of the "politics is glacier" metaphor offers an opportunity for brief comments on a couple of phrases that have a role in a fuller interpretation. First, the lexicon descriptor for "glacier," (see Tab. 6.1 in Section 6.2) "may not be possible to stop," is of dubious salience. However, it does turn out to be relevant for the given context, though dependent on the added complication of relating this descriptor to its *opposite* in the text, namely "still arrestable."

This discursive context also yields an example of metaphors for which perhaps a human or computational paraphrase *cannot* be satisfactory in all its details. The phrase "left behind a rich terminal moraine" does not add to the information supplied by "remarks and tactics," but does provide an entertaining general impression by completing the "glacier" image, while conveying the writer's desire to focus on the abundance of remarks and tactics following momentous events. An interesting point regarding the sometimes secondary nature of logic and information in metaphor is that readers might not even think of the fact that when glaciers leave moraines, they are receding, not advancing. If readers initially held a picture of an advancing glacier in their minds, they might have to change their image to more fully experience the metaphor.

The "ideas are buoys" metaphor is perhaps the most challenging of the examples; the partial nature of the paraphrase is due to the simplicity of the input required by MAP. In addition, the definition of "buoys" in MAP's lexicon contains only the information that it mentally signals something to the viewer of the buoy, since the specific sense of "buoy" as a marker of a boundary or danger is not differentiated. If the participial phrase qualifying "buoys" (i.e., "marking out") is used by the host program to identify the intended sense

3 on the PBS News Hour

of "buoy," then the function of a buoy could be chosen corresponding to the indication of danger, as opposed to quite a few other types and uses of a buoy. With the addition of adjective definitions in MAP's lexicon, the negative connotation attached to "stormy" in the present case would serve as negative reinforcement and the extreme magnitude of "deep" as qualifier of "seas" could be applied to the ideas (the embedded MENTAL INTELLECTUAL concept conveyed by the presence of the buoy), giving:

(IDEAS ARE MEANS OF MENTAL INTELLECTUAL CONCEPT (POSSIBLE (SEQUENCE-OF VERY MANY MENTAL-INTELLECTUAL CONCEPT WITH VERY NEGATIVE CONNOTATION)) GO FROM IDEAS TO USER)

It appears that this metaphor is difficult for a program to paraphrase and perhaps for a human to paraphrase without a lot of elaboration.

The given examples may provide some idea of the problems and possibilities of a computational interpretation of nominal metaphor that goes beyond the introductory nature of MAP. Some metaphors obviously do not fit into the syntactically simple input to MAP's current version. Sometimes an inference on the vehicle nominal may be the actual ground of the metaphor. For the example, "yesterday's conversation is a pebble in my shoe," the inference of the whole phrase "pebble in shoe," i.e., that it "hurts," is what is transferred, namely from the SENSORY-FEEL to the MENTAL-ATTITUDE domain. This kind of inference is not made by MAP, but is a part of the metaphoric-idiom processing program presented in Chapter 7.

Computational metaphor interpretation of nominals, even if lexicon definitions are adequate, may share the NLU problem that words are often carelessly used by speakers/writers, e.g., by treating a prism and a lens as synonyms. Notably, as in the case of verbal metaphor, experience evoked by a metaphor is difficult if not impossible to capture; it is improvements in language representation itself, together with some extension or re-evaluation of abstract components, that presents much of the addressable challenge.

Some of these problems are not of concern for another type of phrase containing metaphorically used nominals, namely metaphoric nominal compounds. For example, these are used by an originator not mainly to evoke a particular way of seeing a concept. However, nominal compounds have interesting complexities of their own, as discussed in the next section.

6.7 Metaphoric nominal compounds

A remaining type of metaphoric phrase consists of two or more nominals taken as a unit. Here we consider the two-nominal form, as we are looking only at the metaphoric aspects of interpretation. A brief analysis of nominal compounds in general precedes the discussion. Many nominal compounds are "frozen" and need no such analysis; however, as in the case of conventional and unconventional metaphor, the procedure is intended to be general, applying to nominal compounds that are and are not lexicalized as a unit. The nonmetaphorical work was developed and implemented in [28]. Finin [11] has also computationally processed nominal compounds, as has Sowa [34], using his "conceptual graphs"; neither appears to have considered metaphoric nominal compounds.[4]

6.7.1 Nominal compounds

In a two–unit nominal compound, the first nominal (referred to here as n1) qualifies the second (n2), but not directly; there is a relationship that links the two together, yielding its disambiguated meaning. Su [36] called this relationship the "interactive meaning" of the compound. For example, "butcher knife," "kitchen knife" and "tomato knife" all differ significantly in their interactive meanings because "knife" is qualified by n1 in a conceptually different way in each; the knife has the function of being "used by," "used in" and "used for" respectively. Many nominal compounds have more candidates for interpretations because of more case relationships. A nominal compound with e.g., "train" as n2, can carry n1 ("coal, commuter"), be located in n1 ("Boston"), have n1 as a source or destination ("Boston"), or be used by n1 ("commuter"), all of which fit into different slots (OBJECT, LOCATION, SOURCE, GOAL, AGENT. In addition, n2 could be owned by n1 ("Boston & Maine Railroad"), have n1 as a part ("[iron–] wheel train"), be physically composed of n1 ("box car") and several other basic relationships.[5] Obviously discursive context and cultural knowledge other than function[6] are often needed for further disambiguation–and ignored for present purposes.

In isolation, the choice of interactive meaning depends on what is conceivable, i.e., how the two nominals can fit together, which itself depends on their characteristics. A "saddlebag knife" is unfamiliar, but it is conceivable because a saddlebag can contain, i.e., is +CONTAIN. While in a sense any combination of nominals is conceivable in some remote way, we can draw a line and reject, for example, "coffee knife."

---

4 in spite of Sowa's use of the term "knowledge soup" as a descriptor of another product of his research! [35]

5 The conceptual–case–based structures used in the original program were in Schank's [31] Conceptual Dependency format.

6 Scripts can provide both expectations of and slots for meanings.

The FUNCTION of an artifact is a cultural, stereotyped property, and, as for other nominal metaphor, is a prime candidate for "salience." Different FUNCTIONs for different nominal combinations, however, depend on conceivability. In each case of n1 above, a knife by itself is defined as used to cut or rather is meant to do so. Filling the various case slots attached to "cut" with different n1s for "knife" as n2, however, yields different interpretations. Aside from familiarity of the nominal compound, we can calculate that a butcher knife is conceivably meant to be used *by* a butcher because a butcher is a person; a kitchen knife is conceivably meant to be used *in* the kitchen, because a kitchen can contain; a tomato knife is conceivably meant to be used to cut tomatoes, because the direct object relationship has high priority and because the other two cases do not apply. One might rather not think about what else a "butcher knife" could mean (but probably does not, because a butcher is a person).

Three tasks are involved in the program to interpret nonmetaphorical nominal compounds: identifying types of interactive meaning; identifying descriptors of the two nominals appropriate for the determination of their relationships; and prioritizing in case of multiple possibilities. The governing links – which qualify n2 in terms of n1 by connecting n2 to the specific interactive meaning involving both n1 and n2 – are themselves of three recognized types. The FUNCTION type has been indicated above. A nonFUNCTION type is either HABITUAL or ACTUAL. Thus while "knife" as n2 likely has a FUNCTIONal relationship with n1, one can imagine a "cupboard knife," in which case the interactive link could be HABITUAL (with an interactive "containment" relationship) or a "steel knife," in which the link could be ACTUAL (composition). Any artifact, as a human-made concept, has high priority as a candidate for a FUNCTION link, while natural concepts, such as "swamp frog," are usually HABITUAL.

As the meaning of a nominal compound depends on the characteristics of each of the nominals, the conceptual/semantic nominal definitions are critical. Definitions of nominals in terms of functions and conceptual features, as described for use in metaphor interpretation, can be used for this purpose, with minor adjustments. Conceptual domains become conceptual features of the nominals. Abstractions are handled as they are for metaphor, e.g. SHAPE applies to MENTAL as well as PHYSICAL. There may be an additional level of information associated with a feature value. +PART and +CONTAIN specify what type the nominal is part of and what type can be contained, respectively.

Expected relationships between n1 and n2 are such that either n1 qualifies n2 directly or n1 is involved in a predicate (often FUNCTION) of n2. An example of the former is "steel knife," interpreted as a knife composed of steel, because n2 is +SHAPE and n1 is – SHAPE, i.e., is a nondiscrete material. An example of the latter is "tomato knife," interpreted as "knife with the function of cutting tomatoes," because the FUNCTION of n2 has an OBJECT slot which specifies a +SHAPE object, which is satisfied by "tomato." The procedure can be enhanced by allowing "usualness" in terms of one of a limited number of categories, in this case that one typical use is for cutting *food*.

Relative priority in the case of multiple conceivable interpretations depends on the "binding strength" of the candidate for the interactive meaning. This binding strength is based on several factors, of which two have been implemented:

1) Priority of the *type* of meaning. For example, the PART relationship, which has very narrow requirements on the characteristics of the noun which represents the part and the noun which represents the possessor, (as in [18–] wheel truck) has a higher priority than CONTAINment.
2) Priority resulting from the satisfaction of certain semantic descriptors which are marked in the dictionary as satisfying a *PREFERRED* binding. If n1 satisfies the PREF criterion, the resulting interpretation will have a higher priority than if it had satisfied only the more general specification.

Priority of interpretation (order of output interpretations) is reflected in the order of application the tests for particular interactive meanings. For instance, the PART test is applied before LOCATION or CONTAINMENT, because the PART relationship specifies what object or type of object the concept is part of. The lack of any result would indicate a possible false parse. The implementation of Finin [11] assigns priority in terms of a score, which is potentially more accurate, because the described application of tests is simply linear, not allowing for interaction between criteria.

Output can show interpretations that are closely related, as for "wine glass":

glass which is a means of drinking wine
glass from which one drinks
glass which contains wine

Other nominal compounds may have very different possible interactive meanings, as for "flower knife":

knife which is a means of cutting flowers
knife which has (images of) flowers on it

For "prize–winner book":

book about prize–winners
book for prize–winners (or meant to be read by prize–winners)
book owned by prize–winner

6.7.2 Application to metaphoric nominal compounds

An implemented system of Mizuguchi and Yamamoto [23] interprets metaphoric nominal compounds, but is focused mainly on what is here called ACTUAL interactive meanings based on simple relationships, such as metaphoric comparison. Many metaphoric nominal compounds are more complicated. The nominal n1, as in the case of nonmetaphoric nominal compounds, may be a simple concept, but may be something that fits into a more complex scenario or predicate, which must itself be "unraveled" in order to determine the interactive meaning of the nominal compound. Some (unimplemented) examples illustrate how an interpretation, or more likely, several interpretations, can be derived for metaphoric nominal compounds on the basis of the procedure for literal nominal compounds. A relatively simple example is:

- idea supermarket

The function of a supermarket from the perspective of most people is to obtain a variety of PHYSICAL items from it; i.e., a supermarket is a SOURCE. In terms of the structural components of MAP, we have:

AGENT (=GOAL)
(OBJECT:HIGH–IN–NUMBER AND MIXED–IN–HOMOGENEITY
AT LOCATION:AGENT) (FROM:supermarket) (TO:AGENT)

in the PHYSICAL domain.[7] "Idea," however, is in the MENTAL domain. As in the case of verbal metaphor, the domain inconsistency between OBJECT and predicate (verbal concept) must be made consistent. Since "idea" in the most likely interpretation (see Section 6.7.1) would be the OBJECT in the abstract structure of "supermarket," it determines the conceptual domain, and the predicate of "supermarket" as SOURCE is extended to the MENTAL domain. Therefore, an interpretation of the example, "The university (company, etc.) is an idea supermarket," is that the university is a SOURCE of many various ideas. (Note that "university" and "supermarket" as institutions have a MENTAL component, which, consistent with "idea," is the relevant domain here.)

7 While the whole transaction between customer and supermarket involves a (CONTROL–domain) "sell" structure (for the supermarket) as well as a "buy" structure (for the consumer), as formulated in Conceptual Dependency representations [31], the nominal compound constrains the focus to the transferred OBJECT, which literally is PHYSICAL.

- White–House key

There are several possible interpretations for this phrase, aside from literal ones. A literal (PHYSICAL) White–House key could be a means of physically entering the White House. Again the functional representation of n2 is significant. The function of a key is to open something. The literal verb "open" can be represented as "to create the possibility of going into or out of." The key can be used to enter the White House or can be (metonymically) used *by* the "White House" to open or close something else. For the former interpretation, based on the structural components of MAP, we have:

AGENT (=OBJECT)
(OBJECT:AGENT AT LOCATION:White–House)

Metaphorically, with the White House metonymically representing the presidency, which is in the CONTROL domain, the structural representation for "open" can be extended to that domain. Thus for "Money is the White–House key," "White–House key" for one interpretation is represented as:

AGENT (=OBJECT)
(OBJECT:AGENT AT LOCATION:presidency).

which can be directly paraphrased as "Money is a means of arriving at the presidency." However, since humans are the LOCATIONs for attributes (see Chapter 5), this representation can also be paraphrased with the OBJECT and LOCATION reversed. Thus just as "The country pranced to prosperity" can be interpreted as "The country [easily] became prosperous," the preceding paraphrase can be converted to "Money is a means of *becoming president*."

- crime pill

This nominal compound (intentionally unfamiliar, as are the subsequent examples) also has an interpretation based on the FUNCTION of n2, namely "pill," which is either to counteract (eliminate or reduce) or enhance some condition of animate beings. As crime has a negative connotation, the function of counteracting is chosen, here represented in a simplified way[8] as eliminating:

AGENT <structure representing use of pill>

---

8 "Disease of +ANIMATE–being" in the PHYSICAL–ANIMATE domain could be expanded abstractly as (NOT (OBJECT:+ANIMATE–being BE WHOLE))

(NOT (OBJECT:disease BE))

With the extension from the PHYSICAL–ANIMATE domain to the EXTRINSIC–CONTROL (of action) domain of "crime," the representation is:

AGENT <structure for doing something>
NOT (OBJECT:crime BE), i.e., a means of eliminating crime

Thus the example, "There is no crime pill" could be paraphrased as "There is no means of stopping the existence of crime."

– information curtain

One function of a curtain is to hide something, a SENSORY–SIGHT concept. It is thus a means of:

AGENT <structure for doing something>[9]
NOT (OBJECT:image AT LOCATION:+ANIMATE–being)

The structure in the MENTAL–INTELLECTUAL domain of "information" is:

AGENT <structure for doing something>
NOT (OBJECT:information AT LOCATION:+ANIMATE–being)

An information curtain would then, in paraphrase, be a means of causing someone to not have information.

However, another possible function of a curtain is to prevent entering or exiting by a concept on one side or the other (a shower curtain prevents water from exiting). In the case that a human cannot enter, the curtain is a means of causing:

AGENT <structure for doing something>
NOT (OBJECT:human AT LOCATION:information)

which, with the reversal due to the human being the OBJECT, is actually equivalent to the preceding representation (which makes sense, since if information is hidden, it cannot be accessed).

If, on the other hand, the *information cannot exit*, we again end up with the same representation. Entering and exiting could of course be included in the same "OR–ed" representations of leaving and entering a state, but the result would be similar.

9 A curtain might also be thought of without an AGENT, i.e., simply as an existing barrier.

A somewhat different case is given by:

- propaganda curtain

The FUNCTION of n2, "curtain," is the same as in the previous example, but may have more interpretations than those corresponding to that example. A "propaganda curtain" is conceivable in the sense of hiding or providing a barrier to propaganda, though somewhat opaque. However, the function of propaganda, like that of a curtain, could *also* be expressed in terms of hiding, i.e., of hiding the truth. The fact that both n1 and n2 have the same function fits the criteria for another test for an interactive meaning, namely COMPOSITION, meaning n2 is composed of n1. The test minimally requires that n1 be−SHAPE and n2 be +SHAPE, as in "steel knife" or "granite pebble." As these features apply in other domains as well, we can obtain the alternative interpretation for "propaganda curtain" as produced by the COMPOSITION test as:

INSTANCE COMPOSED−OF propaganda

in other words, a "curtain of propaganda."

To satisfy the test, the functions of "propaganda" and "curtain" have to be defined in exactly the same way for a COMPOSITION match to occur. Simple abstract representations are an advantage here, in that they are more likely to match than words, such as "hide" for one nominal and "conceal" for the other.

As the two nominals in a nominal compound place constraints on each other, the sense of one or both of them may often be disambiguated in the process of interpretation. For example, for a "(2−) hour class," since "hour" is a TIME concept and a class which one teaches has a temporal component, the sense of "class" is most likely not that of a social group.

A great many nominal compounds, including metaphoric nominal compounds, can thus be interpreted, but usually with a number of candidates for the *intended* interpretation. It seems that the discursive context should play a greater role than at least some of the criteria for priority of nominal compound interpretation. Ultimately, the interpretation of some nominal compounds will be partly subject to the way originators, especially in casual discourse, want to connect the nominals, on the basis of the current situation and past events.

References

[1] Adams, S.: Dilbert. comic strip, *United Media Syndicate* (2002)
[2] Adamson, M.: Graduate education is the dubai of higher learning. http://www.aaup.org/article/graduate-education-dubai-higher-learning#.VEk9vVupU9c (2010). Accessed 10/23/14

[3] Aiden, E., Michel, J.B.: *Uncharted: Big Data as a Lens on Human Culture*. Riverhead, New York (2013)
[4] Bonvin, J.M.: Une operation miracle pour le bétail. Newspaper "La Suisse," Geneva, 25-4-82 (1982). Discontinued 1994
[5] Carbonell, J.: Metaphor: An inescapable phenomenon in natural-language comprehension. In: W. Lehnert, M. Ringle (eds.) *Stragegies for Natural Language Processing*, pp. 415–434. Lawrence Erlbaum, Hillsdale, NJ (1982)
[6] Carbonell, J., Minton, S.: *Metaphor and Common-Sense Reasoning*, Rep. No. CMU-CS-83–110. Carnegie-Mellon University, Pittsburgh, PA (1983)
[7] Courteau, D.: The poetic subversion of capitalism. *Johns Hopkins Magazine* pp. 20–21 (Summer 2011)
[8] Drezner, D.W.: Why wikileaks is bad for scholars. http://chronicle.com/article/Why-WikiLeaks-Is-Bad-for/125628 (2010). Dec. 5, accessed 27-10-14
[9] Fass, D., Wilks, Y.: Preference semantics, ill-formedness, and metaphor. *American Journal of Computational Linguistics* **9**, 178–187 (1983)
[10] Ferris, L.: Of periods, serifs and politics. http://chronicle.com/blogs/linguafranca/2012/08/08/of-periods-serifs-and-politics/ (2012). Accessed 27-10-14
[11] Finin, T.: The semantic interpretation of nominal compounds. In: *1st Annual National Conference on Artificial Intelligence*, AAAI–80, August 1. AAAI Press (1980)
[12] Gibson, J.: The theory of affordances. In: R. Shaw, J. Bransford (eds.) *Perceiving, Acting and Knowing*. Lawrence Erlbaum Associates, Hillsdale, NJ (1977) pp. 67–82
[13] Henahan, D.: The riddle of john cage. *New York Times*, p. D-17 (1981). 23-8-81
[14] Horsely, S.: A closer look at past tax overhaul attempts. http://www.npr.org/templates/story.php/?storyId=129754766 (2010). Aired 9-9-10 on "In Focus," accessed 27-10-14
[15] Janssens, Y.: Une cathedrale de cristal en californie. *La Suisse* (1981). 9-8-81
[16] Johnson, M., Malgady, R.: Toward a perceptual theory of metaphor comprehension. In: R. Honeck, R. Hoffman (eds.) *Cognition and Figurative Language*. Lawrence Erlbaum Associates, Hillsdale, NJ (1980) pp. 259–282
[17] Kajewski, B.: The 2011 mind-set of faculty (born before 1980). http://chronicle.com/article/The-2011-Mind-Set-of-Faculty/128705/ (2011). Accessed 27-10-14
[18] Kaplan, R., Kaplan, E.: *Hidden Harmonies: The Lives and Times of the Pythagorean Theorem*. Bloomsbury Press, New York (2011)
[19] Kilpatrick, P.: An a-frame model for metaphor. In: *Proceedings of the International Conference on Cybernetics and Society*, pp. 83–87 (1982)
[20] McKibben, B.: *Eaarth*. Macmillan, New York (2010)
[21] Miller, G.: Images and models, similes and metaphors. In: A. Ortony (ed.) *Metaphor and Thought*. Cambridge University Press, Cambridge, England (1979) pp. 202–250.
[22] Mirsky, S.: Short takes. *Scientific American* **294**, 102 (2006). In Anti-Gravity
[23] Mizuguchi, F., Yamamoto, A.: An approach to a semantic analysis of metaphor. In: *Proceedings of the 8th International Conference on Computational Linguistics (COLING–80)*, Tokyo, Japan, pp. 136–143 (1980)
[24] Morgan, J.: Observations on the pragmatics of metaphor. In: A. Ortony (ed.) *Metaphor and Thought*. Cambridge University Press, Cambridge, England (1979) pp. 136–147
[25] Ortony, A.: Beyond literal similarity. *Psychological Review* **86**, 161–180 (1979)
[26] Rieger, C.: Conceptual memory and inference. In: R. Schank (ed.) *Conceptual Information Processing*. North Holland, Amsterdam (1975) pp. 157–288

[27] Rosenberg, R.: The silicon valley: A fascinating saga. *Boston Globe, Books–1* (1982). 3-10-82 Review of D. Hanson, *The New Alchemists: Silicon Valley and the Microelectronics Revolution*, Little Brown and Co., Boston (1982)

[28] Russell, S.W.: *Computer Understanding of Conceptually complex Phrases*. Ph.D. thesis, Stanford University (1975)

[29] Russell, S.W.: Computer Understanding of metaphorically used verbs. *American Journal of Computational Linguistics* **Microfiche 44** (1976)

[30] Russell, S.W.: Formalization factors in metaphorical extension. In: *Proceedings of the European Conference on Artificial Intelligence*, pp. 234–239 (1982)

[31] Schank, R.: *Conceptual Information Processing*. North Holland, Amsterdam (1975)

[32] Searle, J.R.: Metaphor. In: A. Ortony (ed.) *Metaphor and Thought*. Cambridge University Press, London (1979) pp. 92–123

[33] Sorren, M.: "The mouse that roared:" how disney instills greed and consumerism–starting at three months. http://http://www.truth-out.org/opinion/item/2138:the-mouse-that-roared-how-disney-instills-greed-and-consumer\ism-starting-at-three-months (2011). Review of H. Giroux and G. Pollock, *"The Mouse That Roared: Disney and the End of Innocence,"* The Rowman and Littlefield Publishing Group, Lanham, MD (2010). Accessed 31-10-14

[34] Sowa, J.: *Conceptual Structures: Information Processing in Mind and Machine*. Addison–Wesley Longman Publishing, Boston, MA (1984)

[35] Sowa, J.: Representing knowledge soup in language and logic. url http://www.jfsow.com/talks/souprepr/htm (2002). *Talk presented at the Conference on Knowledge and Logic*, Technische Universitaet Darmstadt, June 15. Accessed 22-6-15

[36] Su, S.: *A Semantic Theory Based on Interactive Meaning*. Computer Science Technical Report 68, University of Wisconsin, Madison, WI (1968)

[37] Taylor, M.: Graduate education is the detroit of higher learning. http://www.nytimes.com/2009/04/27/opinion/27taylor.html (2009). Accessed 23-10-14

[38] Thaler, R.H.: Breadwinning wives and nervous husbands. *New York Times, Business*, 2 June 2013 (2013)

[39] Vinocur, J.: In bonn, the coalition's decline accelerates. *The New York Times, The Week in Review* p. 2 (1982)

[40] Weiner, E.J.: A knowledge representation approach to understanding metaphors. *Computational Linguistics* **10**, 1–14 (1984)

[41] Weissman, R.: none. http://publiccitizen.org (2014). Accessed 18-1-14

[42] Weizenbaum, J.: *Computer Power and Human Reason: From Judgment to Calculation*. W. H. Freeman, San Francisco (1976)

[43] Winston, P.: Learning by creating and justifying transfer frames. *Artificial Intelligence* **10**, 147–172 (1978)

[44] Wong, K.: Twilight of the neandertals. *Scientific American* **301**, 32–37 (2009). Citing Leslie Aiello.

7 Metaphoric Idioms

Idioms are very common in natural language. Good introductions to idioms are provided by Dobrovolski [6] and van der Linden [16], who includes positive definitions of idiomatic expressions. In this discussion[1] the focus is solely on *metaphoric* idiomatic expressions, and the application of a program such as MAP to interpret those that are modified in some other than a syntactically simple way. In contrast to the part–sentence metaphors that were the topic in the preceding chapters, the idioms considered here are metaphoric as a whole. There is thus no domain inconsistency. Rather, the basic form of the idiom is recognized through its presence as a "big word" in the conventional lexicon, if all the obligatory parts of the idiom are present and fulfill certain syntactic and semantic requirements.

Whether an idiom as a whole is a dead metaphor, such as "stir up a hornets' nest"; has a speculative historical basis, such as "kick the bucket" (as a reference to pigs carried to market upside-down with their feet "kicking" the beam known as a "bucket"); or doesn't seem to make sense at all, such as "cold as a bear," it can be represented as a stock phrase in the lexicon with a nonidiomatic meaning.

Metaphoric idioms are generally decomposable. "Decomposable" here does not mean that an individual part of the idiom has the same meaning outside the idiom–obviously "cat" in the idiom "let the cat out of the bag" does not (though see the reference to puns below). However, while the idiom has a meaning as a whole unit, one or more parts of the idiom can often be modified–in either the vehicle or topic domain, or both–to create a variation based on the meaning of the unmodified idiom.

Only the first of the idioms mentioned above, "stir up a hornets' nest," is considered decomposable. (Ernst [9], however, cites the example, "With his dumb remark at the party, he kicked the social bucket," which can be seen as forcing an adverbial concept modifying the entire idiom into the position of an adjectival topic–domain modifier of "bucket," where "bucket" itself makes no sense but the idiom can be understood.)

Consider the following basic idioms:

let the cat out of the bag (reveal a secret)
take the wind out of [one's] sails (take away [one's] motivation or progress)
rub salt into [one's] wounds (make [one's] bad situation worse)
crack a hard nut/a hard nut to crack (a difficult problem)

---

1 based in part on research with Ingrid Fischer and Ricarda Dormeyer

Examples of variations:[2]

The cat hopped completely out of the bag
...blows fresh wind into the slack sails of the church
They are always rubbing salt and pepper into their open wounds
A hard nut is lying in the basket of the European Union

Creative possibilities appear to be unlimited. To take the example "let the cat out of the bag," the following further variations have occurred:

The education director did not yet let the cat out of the bag, but [did let out] at least the head
He lets the cat at least peek out of the bag
The cat unfortunately escaped out of the bag
However, the public only got to see the tail [of the cat]
He let the cat partly out of the Christmas grab bag
she lets the ideological cat out of the brocaded bag of prose
He lets the cat out of the bag, which is supposed to eat the bad rats

Such variations go beyond those consisting simply of syntactic changes [27] ("The cat was let out of the bag"); quantification ("He let three cats out of the bag"); unmodified reference to the "result" portion of an idiom ("The cat is out of the bag" [8, 20, 1, 7]); and other modifications considered applicable to both vehicle and topic domains ("He has not yet let the cat out of the bag").

A candidate for a metaphoric idiom may of course alternatively have a literal interpretation; a cat may mercifully be literally let out of a bag. In some cases, determination of whether the expression is literal or metaphoric requires reference to its context; however, psycholinguistic research [13] has shown a strong preference for the idiomatic interpretation. Simultaneous metaphoric and literal interpretations are often the basis of puns, as in the case that a cat as a Christmas gift is accidentally revealed beforehand by coming out of a bag (apologies to cat owners). While it is tempting to assert that the metaphorical cat (the secret) is also the literal cat here, it is more accurate to say that the secret is the *identity of the gift*. However, another pun is an unusual case in which "the [metaphoric] bag" is simultaneously close to being literal: "On Thursday the Bern Zoo Daehlhoelzli let a special cat out of the bag: ...a leopard, the offspring of parents Saida and Rigo."[3]

---

2 These two groups of variations are (translations) taken from the COSMAS corpus at the University of Mannheim, Germany www.idsmannheim.de/kt/cosmas.html.
3 ibid.

There are many ways to modify idioms; attempts to interpret them can present interesting and open-ended challenges, as indicated in the remainder of this chapter.

7.1 Basic and modified idioms

Human processing of idioms [3, 4, 10] and its role in defining a process model of idiom parsing (see for example Lytinen et al. [17]) is not addressed here, though the literature suggests that human processing of creatively modified idioms differs somewhat from human processing of familiar idiomatic expressions. According to interdisciplinary studies, however, many idioms are compositional and can be analyzed, and they are interpreted by humans through metaphoric extension from vehicle to topic [18, 14, 12].

Computational recognition of idioms has received much interdisciplinary attention [4] [20]. See for example van der Linden [16] for a workable method. The processing of modified idioms through word changes or additions has been the subject of work by van der Linden [15], Cignoni and Coffey [5], Riehemann [22] and Everaert et al. [10], among others. The metaphor–analytic approach is consistent with psycholinguistic evidence that the literal meanings of at least some of the words of a decomposable idiom and of its variations are referred to in its interpretation [19, 2, 27].

7.1.1 Tasks

As in the case of metaphor based on a causative action, it is usually the result portion of the structure that is of primary interest and receives the domain specification. As noted earlier, a common variation of a basic idiom with a causative verb is to state just the result, as in "The cat is out of the bag." There may also be a critical inference on that result portion or on the meaning of an idiom based on a noncausative action. For the idiom "let the cat out of the bag," the image of the cat in the bag corresponds to the idea that the cat (the mental object) is hidden; if it is let out of the bag, it can be seen by those other than the possessor of the secret. The inference on the presence of the cat outside the bag is that people can see it. Since we know what the general interpretation of the basic form of an idiom is, we can enter any inference which generates this interpretation into the extensible representation of the idiom.

When an idiom in a text is parsed (not a MAP procedure), it must be ensured that all parts of the idiom are found, so that the idiomatic reading can be used. The minimal assumption is that, from the representation of the idiom and its inference, key nominal(s) and an appropriate verb or locative copula must be present. For "let the cat out of the bag" and for the presence of only the result portion, as in "The cat is out of the bag," the nominals are the OBJECT and LOCATION "cat" and "bag" respectively. For the case of the result portion of the idiom, a match with the result portion of the full idiom entry can be attempted, or that portion may have a separate lexical entry.

An appropriate verb in the case of a verb substitution of the verb "let out" must similarly express a locative relationship. It might be observed that the verb in some idioms may never or almost never be replaced, except perhaps in the case of a nonnative speaker who recalls the gist of an idiom but not the actual words. "Spill the beans" might fall into this category; a verb substitution (e.g., "pour the beans,") would place great weight on a nonidiomatic interpretation. There is a fine line, however, between a creatively modified idiom and either a nonidiomatic construction or a mistake.

An interesting though uncommon case is one in which, given that the appropriate nominals are in place, the structure of a verb which is substituted has a disambiguating role. For example, German has not only a direct translation of this idiom, "die Katze aus dem Sack lassen," but also "eine Katze im Sack kaufen" ("to buy a cat in the bag"), corresponding to the English "to buy a pig in a poke," meaning "to buy" (or "to buy into") something without first ascertaining its value. The requisite "Katze (cat)" and "Sack (bag)" are present in this idiom as well. (These two idioms are speculated by some to be related in origin; in medieval markets a merchant ostensibly placing a sold pig into a bag might *secretly* substitute a cat, which he or she then hopes will not exit the bag before the deceived buyer is out of sight. [26])

Nevertheless the two idioms are of course distinct. If the usual words "let out of" or "kaufen/buy" are used, there is no problem in identifying the idiom. If the verb "buy/kaufen" or "let out of/auslassen" is varied in a comprehensible way, there will still be a structural difference between the two verbs. "Kaufen/buy" or similarly structured verbal concepts, while involving a transfer, do not themselves express a transfer out of the bag. Moreover, "buy" is a concept of possession, i.e., in the EXTRINSIC CONTROL rather than the PHYSICAL domain. While a partial match with more than one idiom is infrequent, conceptual-domain information could in some cases aid disambiguation. With these observations, then, a distinction between the meanings of the two idioms, even if modified, could be determined.[4]

4 With quasi-infinite possibilities of situations, however, a scenario can be imagined in which prior discourse or conversation using the idiom "die Katze aus dem Sack lassen" ("let the cat out

The idiom "let the cat out of the bag," a typical decomposable idiom, has parts that may be modified to vary the image called up by the basic idiom. The interpretation of its variations, further discussed in the remainder of this chapter, can be based on that of the basic idiom and the inference found in its representation. There is only one inference, in contrast to the many possible inferences on a nonidiomatic reading. Consistent with the inference postulates of Pulman [20], the "cat" being "in the bag," for example, is directly linked to a fact being secret (not known to others), and a "cat" being "out of the bag" is linked to a fact being known to others. Negation of the result ("The cat is still in the bag") is also quite simple. Both of these modifications are in contrast to modifications that modify the representation structure, as in "the cat peeked out of the bag," which are handled flexibly, i.e., according to semantic extensions of the verb.

One of the more interesting and challenging variations is that a nominal of the idiom is modified, e.g., by a preceding adjectival form. The modifier may be in either the vehicle or the topic domain. In the idiom variation, "The agency decided to take the project under its federally-funded wing," "federally-funded" is in the topic domain and needs no metaphoric processing [11]. In the example, "She let the cat out of the brocaded bag," however, "brocaded" is in the vehicle domain – the domain of "bag." The extensible, abstract representation of "brocaded" must be incorporated into the metaphoric meaning of the idiomatic usage. This type of variation is discussed in Section 7.3.

7.1.2 MAP as applied to idiom interpretation

MAP, with its abstract lexicon, offers a method for addressing simple and richer modifications of idioms through its putative cognitively based process of domain transfer. MAP's approach has been used in the work of Dormeyer, Fischer and Russell [7], which implemented interpretations of examples of basic idioms. In Russell [25], interpretation of idioms with a vehicle-domain modifier as supported by abstract representations is outlined. That approach is extended here.

of the bag") is followed by a declaration of the intention "die Katze im Sack zu kaufen" ("to buy the cat in the bag"), where the "Katze/cat" here is the cat mentioned in the previous utterance. In other words, one offers money to know the secret – a reference to revealing a secret rather than to buying something unknown. This would be an inventive sequence, in which the originator of the idiomatic language forces "cat" to have its idiomatic meaning in a particular subsequent expression.

The extensible parts of MAP's verb definitions relevant to our discussion are:

1. Abstract structures for STATEs, with or without a causative component.
2. Embedded evaluative or emotional effects on particular entities, whether part of the definition of an attribute, such as NEGATIVE for pain, or a subjective connotation, such as POSITIVE for "fresh/breezy" and NEGATIVE for "drafty."
3. A small set of conceptual/abstract features applicable to both literal and metaphorical usages.

Because of the perceived correspondence between literal and metaphoric interpretations of idioms, the meaning *representations* of idioms and their components, regardless of notation, are the most critical part of the interpretation. Since in many cases idioms have a unique topic domain, each idiomatic entry in the lexicon can be provided with both its literal vehicle domain and its topic domain, as well as the mapping between them. (In case an idiom can metaphorically refer to more than one domain, these domains–or indication of a *general* applicability–would be entered into the representation; the actual choice of domain would be subject to the context.)

7.2 Representations

Idioms and their inferences are represented in an idiom lexicon in a way which allows reference to EVENTs and various indexed STATEs within the idiom and inference parts. Previous work [7] implemented a model of the idioms *let the cat out of the bag* and *rub salt into someone's wound* through the use of feature structure representations. That general framework is used here, but illustrated with the abstract representation language [23, 24] described for metaphor, adapted for readability.

7.2.1 The cat in the bag

The idiom "<agent> let the cat out of the bag" is represented in Fig. 7.1. (As usual, articles and other grammatical items are omitted as irrelevant to the example.)

In the "metaphoric inference" portion of the figure, there is more explicit information as to where the secret information is (indicated parenthetically) or stays. This is because mental concepts, in contrast to physical concepts such as cats, do not "leave" one person when they are communicated to another. It can be seen that the event may happen without an animate agent, as in the case of a scientific discovery, for example.

The vehicle and topic designations represent the extension from seeing the cat to knowing the revealed proposition, respectively. More specifically, in the SENSORY–SIGHT domain, OBJ is a view and LOC is the sight faculty of the indicated human; in the MENTAL–INTELLECTUAL domain, OBJ is what is thought or known and LOC is the intellectual faculty/mind.[5] This domain transfer is explicit in the representation, though in practice only the final domain need be indicated. The relative TIME representation follows the format of Reichenbach [21], where "<" represents "temporally before" and "EVENT" represents the time of the represented event.

Literal representation

STRUCTURE

EVENT: STATEb CHANGE–TO STATEe

CAUSE: (AGENT +) (VOL +| –)

  OR (AGENT –)

  (If (animate) AGENT,

  then VOLUNTARY or NOT)

DOMAIN: PHYSICAL

STATEb: AT:in (OBJ:cat_i LOC:bag_i)

STATEe: NOT AT (OBJ:cat_i LOC:bag_i)

TIME: STATEb < EVENT < STATEe

Metaphoric inference

STRUCTURE

EVENT: STATEb CHANGE–TO STATEe

CAUSE: (AGENT +) (VOL +| –)

  OR (AGENT –)

VEHICLE DOMAIN: SENSORY–SIGHT

TOPIC DOMAIN: MENTAL–INTELLECTUAL

STATEb: (AT (OBJ LOC:$AGENT_i$)

  &) NOT AT (OBJ LOC:$others_i$)

STATEe: AT (OBJ LOC:$AGENT_i$)

  & AT (OBJ LOC:$others_i$)

TIME: STATEb < EVENT < STATEe

Fig. 7.1: Representation for "let the cat out of the bag."

5 Omitted here is that the transmitted knowledge necessarily goes through the CP (conscious processor) and *presumably* to LTM (Long-Term Memory).

No evaluation of attitude (POSITIVE, NEGATIVE) is assigned to this idiom. The determination of an evaluation for idioms can be quite subtle. The idioms "let the cat out of the bag" and "spill the beans," for example, both represent approximately the same concept, but the latter can be perceived as having a NEGATIVE EVALUATION. Digging further into other subtleties, one might say that in the former case but not the latter, people are aware that there is some secret and are waiting for the revelation, but a representation of this condition is not likely to be of value in the interpretation of the surrounding discourse.

The representations used, based on a few simple, formal STATEs, enable a match with expressions that refer either to the whole representation or only to a beginning- or end-STATE, as in "The cat is out of the bag." A paraphrase of the complete representation is roughly "one does something such that the information not known to others becomes known to others," while a paraphrase of the end-STATE might be "the information is now known to others." The time/word "now" is included because the assumption (from the idiom representation as a whole) is always that the information was previously unknown.

An adverb such as "now" illustrates a temporal concept which, along with certain other words or phrases, such as those expressing the viewpoint of a writer/speaker ("He surprisingly let the cat out of the bag") apply to both vehicle and topic domains and so are unaltered in paraphrase. A somewhat similar example, "The cat is still in the bag," could be represented structurally as:

STATEb: AT:in (OBJ:cat LOC:bag)
STATEe: AT:in (OBJ:cat LOC:bag) (or simply STATEb)

with interpretation corresponding to the inferences in Fig. 7.1. (An alternative is to use the representation for "the cat is in the bag" and simply take over the word "still" in the paraphrase.)

The above basic and modified versions of the "cat" idiom are all simple references to the location of the cat/secret. Missing is the modification in which the location change of the cat is reversed, as in "[She wished she could] put the cat back into the bag." The representation of this variation is more complex. "Put back [into]" does not have the same abstract structure as "let [out of];" it involves the assumption of the basic idiom, namely that the cat was first let out of the bag. Its components are represented in Fig. 7.2 and their use can be sketched as follows.

The information we want to include in a paraphrase is to make the information unknown (secret) which has previously been intentionally or unintentionally revealed. The expression contains a locative verb which, however, does not have the abstract structure of "let." Its representation (1) in Fig. 7.2 is therefore retrieved from the abstract lexicon (STATEp refers to a previous state).

The abstract representation is instantiated (2) with “cat” and “bag,” which are indexed to correspond across the relevant events/states.

1 *“put back”in abstract lexicon:*
(AGENT puts OBJ back into LOC)
STRUCTURE
EVENT: CAUSE STATEb CHANGE – TO STATEe
CAUSE: (AGENT +) (VOL +)
DOMAIN: PHYSICAL
STATEb: NOT AT:in (OBJ$_i$ LOC$_j$)
STATEe: AT:in (OBJ$_i$ LOC$_j$)
TIME: STATEb < EVENT < STATEe
Link from OBJ$_i$ to:
DOMAIN: PHYSICAL
STATEp: AT: in (OBJ$_j$ LOC$_j$)
TIME: STATEp < STATEb

2 *Instantiation of “put-back” structure with idiom nominals:*
STRUCTURE
EVENT: STATEb CHANGE – TO STATEe
CAUSE: ((AGENT +) (VOL +))
DOMAIN: PHYSICAL
STATEb: NOT AT (OBJ:cat$_i$ LOC:bag$_j$)
STATEe: AT:in (OBJ:cat$_i$ LOC:bag$_j$)
TIME: STATEb < EVENT < STATEe
Link from OBJ:cat$_i$ to:
DOMAIN: PHYSICAL
STATEp: AT:in (OBJ:cat$_j$ LOC:bag$_j$)
TIME: STATEp < STATEb

3 *Idiom lexicon match:*
STRUCTURE
EVENT: STATEb CHANGE – TO STATEe)
CAUSE: (AGENT +) (VOL +| –)
OR (AGENT –)
DOMAIN: PHYSICAL
STATEb: AT:in (OBJ:cat$_i$ LOC:bag$_j$)
STATEe: NOT AT (OBJ:cat$_i$ LOC:bag$_j$)
TIME: STATEb < EVENT < STATEe

4 *Metaphoric inference of basic idiom:*
STRUCTURE
EVENT: STATEb CHANGE – TO STATEe
CAUSE: ((AGENT +)(VOL +| –))
OR (AGENT –)
VEHICLE DOMAIN: SENSORY – SIGHT
TOPIC DOMAIN: MENTAL– INTELLECTUAL
STATEb: (AT (OBJ LOC: AGENT$_j$)
&) NOT AT (OBJ LOC:others$_j$)
STATEe: AT (OBJ LOC:AGENT$_j$)
& AT (OBJ LOC:others$_j$)
TIME: STATEb < EVENT < STATEe

5 *Metaphoric inference of modified idiom:*
STRUCTURE
EVENT: STATEb CHANGE – TO STATEe)
CAUSE: ((AGENT+) (VOL +))
VEHICLE DOMAIN: SENSORY – SIGHT
TOPIC DOMAIN: MENTAL– INTELLECTUAL
STATEb: (AT (OBJ LOC:AGENT$_j$)
&) AT (OBJ LOC:others$_j$)
STATEe: AT (OBJ LOC:AGENT$_j$)
& NOT AT (OBJ LOC:others$_j$)
TIME: STATEb < EVENT < STATEe
Link from OBJ to:
VEHICLE DOMAIN: SENSORY – SIGHT
TOPIC DOMAIN: MENTAL– INTELLECTUAL
STATEp: AT (OBJ:info$_i$ LOC:AGENT$_j$)
& NOT AT (OBJ:info LOC:others$_j$)
TIME: STATEp < STATEb

Fig. 7.2: Representations for idiomatic interpretation of “put the cat back into the bag”

Since the idiom is defined as depending on the in-ness or out-ness of the cat and the abstract structure of the verb, the program can match the content of the STATEs representing the cat being in or out of the bag in the instantiation of (2) with the corresponding content of STATEs in the basic idiom representation (3). Specifically, STATEp of (2) matches STATEb in (3), the basic idiom, in which STATEe matches STATEb of (2) (cat out of the bag), which has STATEe as the cat put back into the bag. STATEe of the corresponding metaphoric inference (5) then puts the information back into the sole possession of the agent.

Literal representation
STRUCTURE
EVENT: CAUSE STATEe
CAUSE: (AGENT +) (VOL +| –)
OR (AGENT –)
TIME: EVENT < STATEe

VEHICLE DOMAIN: PHYSICAL
STATEe: AT: in (OBJ:salt LOC: wound(IPART:one_i))
TIME: EVENT < STATEe

Metaphoric inference
STRUCTURE
EVENT: CAUSE STATEe
CAUSE: (AGENT +) (VOL +| –)
OR (AGENT –)

VEHICLE DOMAIN: PHYSICAL
TOPIC DOMAIN: MENTAL– ATTITUDE
STATEe: BE (OBJ:feeling(IPART:one_i) (EVALUATION: (NEG HIGH)))
TIME: EVENT < STATEe
(The OBJ role here is independent of the OBJ role for "salt" above)

Fig. 7.3: Representation for "rub/pour salt in[to] one's wound."

A similar process applies to the more interesting variation, "We can't cram the cat back into the bag" [28]. Here the verb "cram [into]" is embellished by the abstract concept FORCE: HIGH, pointing to (a somewhat humorous image of) desperately putting the cat back/undoing the original revelation.

These representations are based on the cat being either in or out of the bag. It might be countered that in "The cat just peeked out of the bag," it is neither in nor out. As the focus is on what is out, however, being partly out of the bag can be represented with the basic template of being out: STATE (NOT AT (OBJ:(IPART:-cat) LOC:bag), where the MAP component IPART means "inalienable part."

For variations of the basic idiom in which there is no explicit agent, a simple adjustment has to be made to specify an event missing this agent. An example[6] is:

"The Molto Allegro then lets the cat peek out of the bag."

Here the missing agent is a cause (see Section 5.2.1.2 in Chapter 5) which if made explicit would be expressed as the event of playing or performing the Molto Allegro.

7.2.2 Salt in the wound

Another basic idiom, which has as a result a change of attribute, is "to rub/pour salt into someone's wound," with the representation as in Fig. 7.3.

The nominal "wound" is an inalienable part (IPART) of someone ("one"), or of "her," "Matt's," etc. as it occurs in the text. For this idiom, as in the case of the verb "let" in the previous example, another verb used in the vehicle domain can be recognized as taking the place of "rub" or "pour," as long as it has a similar abstract structure. Additionally, the salt must have a LOCATION relationship to the wound – "in/into" or "on/onto" for the basic idiom, with other possibilities (e.g., "washed out of") in a creative variation. The beginning state is omitted, since the previous location of the salt is of no interest.

Extension to the topic domain differs from that in the "cat" idiom, in that "wound" to describe a feeling is in itself a recognizable frozen metaphor based on its physical sense. The nominal "wound"is commonly used outside the idiom, in contrast to, e.g., "cat" or "bag" outside its idiomatic use. Consistent with MAP's treatment of frozen metaphors similarly to that of novel metaphors, the nature of the metaphorical extension is explicit in the representation, with MENTAL – ATTITUDE serving as the topic domain.[7]

As in the case of the "cat" idiom, just the result portion of the event may be expressed. The overheard utterance "[More] snow is like salt in the wound" is

6 COSMAS, in translation

7 In a somewhat more complex process, an inference of pain from the presence of a wound could be represented as a metaphoric extension from the SENSORY – FEELING subdomain. Both mechanisms would result in an interpretation of a high negative evaluation in the MENTAL – ATTITUDE domain.

handled similarly to "the cat is out of the bag." This expression is another example of the extension of the connotation of the vehicle to the topic ("more snow," which does not have an objective evaluative connotation) to the opinion of the *speaker*, no doubt tired of shoveling, namely EVALUATION: (NEGATIVE HIGH).

7.3 Creative variations

The preceding modifications depended on simple logical variations, in that they involved the presence, negation or reversal of states. Many modifications add components to the verb while leaving the logical result the same. Variations of verbs, along with vehicle–domain modifiers are arguably the most interesting variations of idioms. For the "cat" idiom we have the verbal variations "peeked–" or "hopped–out of the bag;" for the "salt" idiom, the adjectival modifiers "open–," "half-healed–" or "weeping wound." Novel variations of this kind can be interpreted–more or less adequately–through components that are extensible to the topic domain.

For example, in the case of "The cat hopped out of the bag," the abstract definition of "hop out of" would include (beside the structure representing the change of location) a NON–CONTINUOUS component. A detailed literal description of "hop" in the lexicon, including, e.g., "to propel oneself upward and land again," while perhaps accurate and useful for other purposes, would not necessarily lead to a meaningful interpretation. As in metaphor interpretation, the program arrives at either a word literal for the topic domain with the same components, if there is one, or simply uses the topic domain translation of the individual components themselves. In this case, the idiom variation "The cat hopped out of the bag" becomes "The information that was not known to others suddenly became known to others."

The substitution of a verb or of a present or past participle, i.e., a modifier based on a verb, is usually less challenging than a vehicle–domain modifier, since all its major components can be represented, while the meaning of an adjectival vehicle–domain modifier may depend on the nominal it modifies. Thus a "healing wound" can be represented as the continuous elimination of the negative feeling, because of the verbal character of "healing," while "open wound" is more difficult for MAP in the absence of the specific meaning of the polysemous word "open" as applied to wounds.[8]

---

8 The literal meaning of expressions such as "open wound,"where the adjective represents simply a type of the nominal, can be determined through other NLU methods.

A more detailed discussion of an adjectival vehicle–domain modification illustrates some of the difficulties of capturing metaphoric meanings in putative literal representations. Consider the modified idiom, "On p.269, she lets the ideological cat out of the brocaded bag of prose."[9] This variation rests on an analogical form, where the topic domain concepts ("ideological" and "prose") are explicitly given along with the corresponding vehicle concepts "cat" and "bag." Similarly to the "political glaciers" metaphor being considered as glaciers representing political situations (see Section 6.3.2 in Chapter 6), the "ideological cat" can be viewed as the cat representing ideology for purposes of the idiom.) In contrast to most other variations of the basic idiom, it is the "bag" which receives metaphoric modification. The "container" of the secret according to the idiom modification is therefore "prose" that is *read*, not someone's *mind*.

In the interpretation process, "bag" is mapped to the (MENTAL–INTELLECTUAL) "prose" of the AGENT ("she"), therefore becoming the LOCATION of the "ideology."[10] The revelation of a secret is again is inferred from the key words of the input phrase and from the basic structure. Interpreting the variation, however, requires the analysis of the modifier "brocaded" and its application to the topic sense of "bag," i.e., "prose."

While MAP does not currently process metaphorically used adjectives or past participles such as "brocaded," MAP's representation scheme could be adapted to their interpretation. If adjectives are defined in the lexicon abstractly in terms of conceptual features, then, we can apply those extensible features to the nominal which the adjective modifies. "Brocaded" could be minimally defined in the abstract lexicon as a PHYSICAL–domain concept with the conceptual features (COMPLEX +) and (EVALUATION: POSITIVE).

The interpretation of the modified concept, then, consists of applying the feature values of "brocaded" to the nominal "prose," in the MENTAL–INTELLECTUAL domain. It might be pointed out that the EVALUATION as extended to the topic applies to a *representation* of a MENTAL–INTELLECTUAL concept, not the content. In other words, "brocaded" is about the written *form*, not the content.[11]

If these components are integrated into the basic idiom representation for "let the cat out of the bag," we have the interpretation in Fig. 7.4, resulting in

9 COSMAS

10 There is an ambiguity involving the semantics of the preposition "of," though without consequence as applied to the idiom, in that it is not clear whether "bag of prose" means "a bag made of prose (with the bag/prose containing ideology)" or "a bag containing prose (with prose–and only indirectly the bag–containing ideology)."

11 MAP recognizes such cases by a REPRESENTATION feature for relevant MENTAL–INTELLECTUAL concepts (see Section 5.5.1).

the paraphrase: "She does something such that the ideology in her positive, complex prose becomes known to others."

Literal representation
STRUCTURE
EVENT: STATEb CHANGE–TO STATEe
CAUSE: (AGENT:she) (VOL +| –)
OR (AGENT –)

DOMAIN: PHYSICAL
STATEb: AT:in (OBJ:cat$_i$ LOC:bag$_j$)
STATEe: NOT AT (OBJ:cat$_i$ LOC:bag$_j$)
TIME: STATEb < EVENT < STATEe
Link from bag$_j$ to nominal in abstract definition of "brocaded:"

DOMAIN: PHYSICAL
STATE: BE (bag$_j$ (COMPLEX +) (EVALUATION: POS))

Metaphoric inference
STRUCTURE
EVENT: CAUSE STATEb CHANGE–TO STATEe
CAUSE: (AGENT:she) (VOL +| –)

VEHICLE DOMAIN: SENSORY – SIGHT
TOPIC DOMAIN: MENTAL– INTELLECTUAL
STATEb: AT (OBJ:ideology$_i$ LOC:prose$_j$ (COMPLEX +) (EVALUATION: POS))
& NOT AT (OBJ:ideology$_i$ LOC:others$_j$)
STATEe: AT (OBJ:ideology$_i$ LOC:others$_j$)
TIME: STATEb < EVENT < STATEe

Fig. 7.4: Representation (with instantiation) for "She let the ideological cat out of her brocaded bag of prose."

Ways in which this simple representation is seen as not fully adequate offer an opportunity for some observations. First, a POSITIVE evaluation of "brocade" may not be acceptable to some, but one can assume that the intention behind a literal brocaded product is to make something appealing. The COMPLEX feature accounts for a good part of the meaning of "brocaded," but lacks the quality of intricacy. A SCALE component would aid in capturing the "detail" aspect of "brocade," with a "small" value. This component is related to MAP's SIZE component, is supported by math-like concepts such as ratio, proportion, etc., and is widely used computationally. A fine (small) SCALE can be thought of as a feature of "brocaded" and *may* indicate the way the author writes her prose, i.e., with many small embellishments.

As a second possibility, words representing more complex concepts could be integrated into the representation system, which would convey more than the two or three abstract components can provide. For "brocaded," an example would be "exquisite," a subjective concept which is equally applicable to several domains ("an exquisite cloth, symphony, wine, performance, etc.") Words such as this, representing complexities and connotations difficult to represent systematically, could be accessed in a system which stores such words as attributes of nominals. However, not only would there be other attribute words stored with the nominal, possibly "old–fashioned," but those words themselves would, dictionary–like, have a variety of attributes, with no indication as to which is to be highlighted in a given expression.

A third method would be to enrich the interpretation of a complex idiom by adding some semantic context of the word in question, in the form of adverb–like descriptors to the paraphrase. A clue to what is missing in COMPLEX as a descriptor lies in the fact that "brocaded" is a past participle, referring to "brocading," i.e., an *action* done in the past, which itself brings with it further semantic descriptors. One such descriptor in this case, informally, might be "carefully adding to" (brocade or embroidery is carefully added to a cloth). Its representation could include the structure for "CAUSE to be MORE COMPLEX," together with a component representing great intensity (FORCE:HIGH) in the MENTAL domain. If used to supplement a representation as in Fig. 7.4, this measure would give a richer interpretation, perhaps aiding an image of what the author of the prose is really trying to do.

The first of these suggestions, while falling short of what the originator of the metaphor is trying to say, is of some value. The second option is "correct" for this idiom interpretation, though it may include irrelevant words in the paraphrase and fall into the trap of an opaque, noncanonical definition system. Both kinds of resulting interpretations would appear to leave the meaning of the intention of the author of the prose ambiguous. While the third suggestion adds reliability to an interpretation, it too misses the "impression" given by the whole idiom. The metaphoric variation gives the sense that "brocaded" is used to set itself apart from the topic–domain concept "ideology"–a hardly exquisite concept. A layer of irony is perceptible here that must be captured by an NLU method beyond the scope of this discussion.

Further experimentation with a greater number of creative examples is needed. As with other types of metaphor, however, a level of adequacy that enables the NLU system to extract pertinent information may be a more realistic goal.

References

[1] Cacciari, C.: The place of idioms in a literal and metaphorical world. In: Cacciari and Tabossi [4] (eds.) *Idioms: Processing, Structure, and Interpretation*, pp. 27–56 (1993)

[2] Cacciari, C., Glucksberg, S.: Understanding idiomatic expressions: The contribution of word meanings. In: G.B. Simpson (ed.) *Understanding Word and Sentence*, pp. 217–240. Elsevier, Amsterdam (1990)

[3] Cacciari, C., Tabossi, P.: The comprehension of idioms. *Journal of Memory and Language* **27**, 668–683 (1988)

[4] Cacciari, C., Tabossi, P. (eds.): *Idioms: Processing, Structure, and Interpretation*. Lawrence Erlbaum, Hillsdale (1993)

[5] Cignoni, L., Coffey, S.: A corpus-based study of italian idiomatic phrases: From citation forms to "real-life" occurrences. In: *Proceedings of Eurolex '98*, pp. 291–300 (1998)

[6] Dobrovolski, D.: *Idiome im Mentalen Lexikon*. Wissenschaftlicher Verlag, Trier, Germany (1997)

[7] Dormeyer, R., Fischer, I., Russell, S.W.: A lexicon for metaphors and idioms. In: D. Schnorbusch, S. Langer (eds.) *Semantik im Lexikon, Tübingener Beiträge zur Linguistik*, pp. 205–224. Gunter Narr Verlag, Tübingen (2005)

[8] Erbach, G., Krenn, B.: Idioms and support verb constructions. In: J. Nerbonne, K. Netter, C. Pollard (eds.) *German in Head-Driven Phrase Structure Grammar, no. 46 in Lecture Notes*, pp. 365–396. CSLI Publications, Stanford (1994)

[9] Ernst, T.: Grist for the linguistic mill: Idioms and "extra" adjectives. *Journal of Linguistic Research* **113**, pp. 51–68 (1981)

[10] Everaert, M., van der Linden, E., Schenk, A., Schreuder, R. (eds.): *Idioms: Structural and Psychological Perspectives*. Lawrence Erlbaum Associates, Hillsdale, NJ (1995)

[11] Fischer, I., Keil, M.: Parsing decomposable idioms. In: *Proceedings of the 16th International Conference on Computational Linguistics*, pp. 388–393 (1996)

[12] Geeraerts, D.: Specialization and reinterpretation in idioms. In: Everaert et al. [10] (eds.) *Idioms: Structural and Psychological Perspectives*, p. 5776 (1995)

[13] Gibbs, R.: Spilling the beans on understanding and memory for idioms in conversation. *Memory and Cognition Proceedings of Eurolex* '98 149–156 (1980)

[14] Gibbs, R.: Idiomaticity and human cognition. In: Everaert et al. [10] (eds.) *Idioms: Structural and Psychological Perspectives*, pp. 97–116 (1995)

[15] van der Linden, E.: Idioms and flexible categorical grammar. In: M. Everaert, E. van der Linden (eds.) *Proceedings of the 1st Tilburg Workshop on Idioms*, pp. 127–143. Institute for Language Technology and AI, Tilburg, NL (1989)

[16] van der Linden, E.: Idioms, non-literal language and knowledge representation. *Computational Intelligence* **8**, 433–455 (1992)

[17] Lytinen, S., Burridge, R., Kirtner, J.: The role of literal meaning in the comprehension of non-literal constructions. *Computational Intelligence* **8**, 416–432 (1992)

[18] Nayak, N., Gibbs, R.W.: Conceptual knowledge in the interpretation of idioms. *Journal of Experimental Psychology: General* **119**, 315–330 (1990)

[19] Nunberg, G., Sag, I.A., Wasow, T.: Idioms. *Language* **70** (3), 491–538 (1994)

[20] Pulman, S.: The recognition and interpretation of idioms. In: Cacciari and Tabossi [4] (eds.) *Idioms: Processing, Structure, and Interpretation* (1993)

[21] Reichenbach, H.: *Elements of Symbolic Logic*. MacMillan, London (1947)

[22] Riehemann, S.Z.: *A Constructional Approach to Idioms and Word Formation*. Doctoral Dissertation, Stanford University. Stanford University, Stanford, CA (2001)

[23] Russell, S.W.: Information and experience in metaphor: A perspective from computer analysis. *Metaphor and Symbolic Activity* **1**, 227–270 (1986)

[24] Russell, S.W.: Metaphoric coherence: Distinguishing verbal metaphor from anomaly. Computational Intelligence, *Special Issue on Non-Literal Language* **8**, 553–574 (1992)

[25] Russell, S.W., Fischer, I., Dormeyer, R.: The cat and the brocaded bag: Using metaphor analysis to computationally process creatively modified idioms. In: B. Sharp, M. Zock (eds.) *Natural Language Processing and Cognitive Science (NLPCS 2007)*, pp. 27–37. INSTICC Press, Portugal (2007)

[26] Soniak, M.: What's the origin of "let the cat out of the bag"? url http://mentalfloss.com/article/31180/whats-origin-let-cat-out-bag (2012). Accessed 10-3-15

[27] Stock, O., Slack, J., Ortony, A.: Building castles in the air: Some computational and theoretical issues in idiom comprehension. In: Cacciari and Tabossi [4] (eds.) *Idioms: Processing, Structure, and Interpretation* (1993)

[28] Wainwright, S.: Scott and bailey. *ITV television*, UK (2011). Accessed 14-2-15, MPBN

8 Conclusion: Possibilities and limits

8.1 Summary

The computational task presented in this book has been to produce general paraphrases for isolated part–sentence metaphor on the basis of the semantics of literal interpretations of the metaphorically used words. MAP, the program implementing this task, includes in its scope the interesting form of nominal metaphor, with its challenging necessity to determine what is likely salient to the vehicle concept, as well as metaphoric nominal compounds. In addition, the metaphor analysis provided by MAP supports the interpretation of modified metaphoric idioms. While MAP interprets conventional and unconventional metaphor in the same way, its focus is on novel, cross–modal, similarity–creating metaphor. Paraphrases are in terms of only basic components, but the minimal detail is related to the ambitious expectation that metaphoric input may come from the domains of all the human faculties.

MAP is treated as an NLP module, focused on the metaphor interpretation and paraphrase process itself and leaving aside less relevant tasks that an operational NLP system must include. With this focus, *given* that an expression is assumed to be nonliteral, MAP does not consider the discursive context surrounding the expression to be interpreted. Neither does it disambiguate words as to their senses; at the present stage of development, the metaphoric input is assumed to contain the unique word sense in MAP's lexicon. In any case, a potential word sense can be supplied by the host NLP program.

Through its ability to interpret a large class of verbal metaphor by the "undoing" of "abstract" nominals derived from reification of verbal or attributive concepts, MAP can paraphrase such metaphor in terms of verbs or attributes as an alternative to keeping the nominal in the paraphrase. Theoretically, by finding words with the same abstract structure in a different conceptual domain, MAP can produce a paraphrase which is another metaphor, though such paraphrases will likely seem strange. In the process of examining relationships between metaphorically used verbs and their immediate semantic contexts (i.e., nominals in case relationships with the verbs), metaphor can be distinguished from incoherence, though this line will always be fuzzy because of the inventiveness of human recipients in finding interpretations.

Beside the method of MAP and similar semantics–based computer models of metaphor, there are additional ways of interpreting metaphor in discourse (beyond simply looking up a dead metaphoric sense in the lexicon). In recent years research on metaphor has proceeded through corpus-based approaches, such as by Barnden et al. [2] and others, to improve automatic text processing. Models based on Lakoff and Johnson's conceptual metaphors are productive, but are limited to interpreting fairly conventional metaphor, since conceptual metaphors are focused on direct topic–vehicle links. Except for an "educated guess" from the discursive context, metaphor that is truly novel would seem to be interpretable only the way humans are thought to do it, that is, by using their understanding of a known usage of the metaphoric unit and extending it to the topic as outlined in this book.

The preceding discussion suggests that computer analyses of metaphor can reveal much of what the originator wants to say and possibly why. MAP does this for verbal metaphor through a scheme which includes both denotations, in terms of structural components, such as an effect or resulting state, and connotations which describe attitudes, emotions and typical effects on participants in the metaphoric image and on a recipient of the metaphor. Chapter 7 has shown that the interpretation of some metaphoric idioms also depend on effects in the form of inferences which are a part of their representations. MAP's treatment of nominals used as metaphoric vehicles highlights the role of stereotypes, subjectivity and multiple perspectives in the determination of what the nominal probably means, as illustrated by the "glacier" example in Chapter 6.

The discussion has also pointed out paraphrase difficulties due to language–specific quirks. Much of this problem has to do with the pervasiveness of metonymy or ellipsis. As observed with respect to an example in Chapter 6, information doesn't come from buoys, but rather from *processing the view* of buoys. There is still work to be done involving how such cases should be handled and how they should be interpreted in the topic domain. In other respects, however, it should be relatively easy to build onto a design such as MAP's, with its small set of descriptors, by refining and adding to its relatively simple representations in a way which still leaves them transparent.

8.2 Interdisciplinary pursuits

The value of exploring the semantic relationships operative in metaphor is highlighted by recent studies in neuroscience to determine activation of sensory–motor areas of the brain by metaphors. Input from these studies could shed light on what enables the *reminding* that lies at the root of metaphor, and is relevant to theories of "embodied cognition" as put forth by Lakoff and Johnson [7].

In one experiment, Desai et al. [5] used functional magnetic resonance imaging to observe brain area activation in subjects while they read both literal and metaphorical expressions containing action verbs, where the verb was the vehicle for the metaphor. It turned out that areas of subjects' brains associated with control of action were activated in both cases. Thus a connection was established between brain activity and properties of the vehicle concept as it was used to understand the topic. Experiments by Narayanan [10, 6] also appear to explain the directionality of metaphor; for example, the metaphor "affection is warmth" is supported, but not "warmth is affection."

The results on brain area co-activation, however, may also call up ideas of a kind of "metaphor therapy" or a metaphoric action in which the topic and vehicle are in fact reversed. For example, Shakespeare images "mercy" as falling "like the gentle rain from heaven"–a "mercy as gentle rain" simile, where "mercy" is the topic, evoking "gentle rain" as the vehicle. In an apparent reversal, one could also imagine being in a gentle rain, where the vehicle, "gentle rain," evokes feelings (through a "washing away" of guilt?) related to the topic, i.e., "(needed) mercy." This would appear to be a "gentle rain as mercy" simile or metaphor.

Consistent with this direction, psychological experiments placing subjects in a "vehicle" situation biased their judgments of the topic because of vehicle connotations. For example, a warm situation evoked personal "warmth," sympathy, etc.; heaviness evoked importance. [3, 1, 9] Neurolinguistic experiments, with words rather than situations as input, appear to contradict this reverse effect, as it rejects "warmth is affection." Perhaps there is no inconsistency, however; the apparent reversal of topic and vehicle in the psychological experiments could still be seen as a "use" of the metaphor, e.g., "affection is warmth." As Lakoff [6] says, the metaphor circuit is activated *as a whole.*

Idioms present an interesting case for brain activity experiments. As pointed out in Chapter 7, many if not most idioms are metaphoric so that some brain activity activity in the area representing the vehicle region *might* be expected. The neurolinguistic results obtained by various experimenters have been contradictory, perhaps because some idioms are not transparent as to their origin and most are frozen. Desai et al. confirmed the intuition that the more idiomatic or frozen the idiom, the less the activity in the relevant area of the brain.

8.3 What is missing

A notable omission from MAP's domain span, and presumably from any computational model, is what might be called the "transcendent" or "spiritual" domain, in which some of the most novel metaphors belong. There is a reason why this kind of metaphor is not included (aside from questioning what the computational use could possibly be!). The semantic and pragmatic basis for the representations which support MAP and other metaphor interpretation systems is by necessity conventional in a broad sense, involving socially created language, stereotypes, "folk psychology,"etc. "Spiritual language"–perhaps an oxymoron, or itself a metaphor–is intended to refer to what is transcendent by definition, i.e., beyond the above conventional notions. It seeks to express, typically through metaphor, that which cannot be literally described–a characterization implied in this book to pertain to a lesser extent to other creative metaphors as well.

Much philosophical language and some poetic language, though not always metaphoric, is subject to similar challenges (consider Augustine's "Transcend time in order that you may be"). Only half facetiously, "interpretation" of such phrases may be best (or "at best") pointed to by indicating what they do *not* mean. This kind of language, metaphoric or not, can of course be and is being explored, with different means and aims. But it does not seem amenable to computational paraphrase. Even with this kind of metaphor put aside, however, there appear to be aspects of metaphor, other than metaphors exhibiting only physical similarity, which one cannot count on to be captured in literal interpretations. Reasons can be given from two perspectives.

8.3.1 Translation

The first reason for less than comprehensive paraphrases concerns the nature of representation of the interpretation, which is essentially a translation. If an interpretation is given in terms of a literal paraphrase including specific words not in the original expression, as it is in some computational models, one cannot expect those words to accurately express what is intended, any more than two "synonyms" are usually not truly synonymous. (Consider the slight nuance between the understanding or use of a word in one language which is supposedly equivalent to its translation in another language.) Sometimes there is even *no* good candidate for a literal paraphrase, because of "lexical gaps." Or, as a metaphor often has connotations and implications, these elements may have to be paraphrased with several expressions, breaking up the image evoked by the metaphor.

At the very least, "literal" words paraphrasing cross–modal metaphor are often etymologically metaphoric–not significant for computational purposes, but perhaps another indication of different perceptions of nonphysical concepts. Etymologies that presumably reflect attempts to refer to nonspatial concepts through spatial ones, are sometimes quite different for different languages. For example, even the words for a very common concept, i.e., "under–stand," "ver–stehen" and "com–prendre," are all different metaphors for (presumably) the same concept in English, German and French respectively. Yet "under" in "understand" does not correspond with "ver–" in "verstehen"; "stand" does not correspond with the French "comprendre," which is metaphorically derived from the physical vehicle "take together." One has to assume that the creation of such words in the past represented different metaphoric views, which may hold for words in different cultures today.

Representation of metaphor in terms of primitive components, as in MAP, is a more conservative approach than translation into a specific word(s); it is quite reliable in its minimal interpretation but sacrifices more detailed information. If the paraphrase is given in terms of basic components, the nature of componential analysis itself has its own theoretical limitation as a kind of reductionism. When an analysis of an expression converts it to elementary states, changes of states, etc., the image on which the meaning of the metaphor relies is broken and something is lost. Components of a metaphor may act synergistically or holistically for humans; the relationship between them may be indicated by abstract concepts but not completely conveyed. Therefore we cannot assume that a metaphor can completely be represented through symbolic components – primitive or not. Black's [4] positing of the loss of cognitive content for human explication of metaphor applies with at least as much force to computer analysis.

8.3.2 Experience

A second reason for the relative inadequacy of a literal paraphrase is related to the first. Especially for rich, expressive metaphor, a purpose is experience, and analytically derived chunks of information cannot be fully experienced. The unique nature of an experience would appear to be especially opaque to representation by general or primitive components, because the components are just abstracted from the experience.

Though a paraphrase may be adequate in reporting literal attributes and actions, and in referring to affective connotations, it does not make the attitudes or feelings of any participants in the metaphoric image come alive, and does not provide the amusement that some metaphors do. Neither is a literal paraphrase as good as a metaphor in conveying other human qualities, such as a sense of wonder. The more expressive the metaphor, even a simple one such as "Juliet is the sun," the more a computer output analysis seems to be lacking.

Details only implicit in the metaphor can provide much of the interest particular to the metaphor. If we consider the "political situations are glaciers" metaphor in Chapter 6, we can see that basic information can be given in a literal paraphrase; however, the richness of the detail given in the metaphor, the caricature of political observers as scientists taking fine measurements of the resulting physical chaos would likely be missing. This richness would also be diminished if the recipient has never experienced the concept represented by the vehicle, i.e., for the "glacier" example, has never seen the debris left in a glacial moraine. The same would apply to "program music" as a metaphor for, e.g., an experience of nature or a narration of adventures.

Thus a computer analysis of metaphor cannot re-present all that *could* be conveyed as a result of the originator's intention – the experience which unites the recipient with the originator's perception, which entertains or manipulates, which offers different highlighting for different individuals. Experience, understanding, and meaning are obviously different from *description* or the symbols drafted to represent them; paraphrases of metaphor which are not themselves evocative of experience are useful as conveyors of information but are not equivalent to the metaphor.

Similar difficulties exist in explaining ostensibly nonmetaphoric language that humans apparently understand, along with the humor. Regarding the music of Richard Wagner, Mark Twain quipped, "It's better than it sounds." Given some thought, an expression of this type could probably be roughly explained by humans and incorporated into a computer model. For example, "better" is relative to a quality in some domain; perhaps a domain other than that of "sound," e.g., music theory. However, this "explanation" does not capture the "Twain-like" humor, along with other nuances of attitude. In any case, there is a quasi – infinite number of different kinds of opaque expressions, not "known" by the system before they are encountered. Many thoughts are not phrased correctly; consider "I spent plus or minus six years in Hong Kong." And why is "a bug on the windshield of life" slightly humorous? Each case might be able to be handled in some way, but there will always be different kinds of cases. Such observations apply all the more to creative metaphoric language.

Similar limitations apply, however, to human "translations" of metaphor in terms of a literal sequence of words. As MacKay [8] indicates, meaning is not always what the individual is able to say it is. Paul [12] appears to go further, saying that explanations of metaphors seem to be inadequate only because we do not really know what they mean. Or, the reason for the inadequacy of an interpretation may be that literal description of a conveyed experience is inherently inadequate simply because it is description. Consistent with Searle's [13] conclusion that a paraphrase cannot reproduce the semantic content of the intended metaphoric meaning, this may define the limitations of human paraphrase of metaphor, and is a theoretical limitation for computer programs as well.

However, progress remains to be made in semantics-based computational efforts to interpret metaphor. The scope of linguistic metaphor is wide and the exploration of metaphor fascinating. In the meantime, we can recall that as Panikkar [11] expresses–metaphorically, (p. 15) "Any *meta–phora* 'carries us beyond' the first meaning of a sentence and allows us to land on unknown shores–and perhaps to get lost in them."

References

[1] Angier, N.: Abstract thoughts? the body takes them literally. url http://www.nytimes.com/2010-02-02/science/02angier.html (2010). Accessed 17-6-15

[2] Barnden, J., Lee, M., Littlemore, J., Moon, R., Phillip, G., Wallington, A.: *Corpus-based Approaches to Figurative Language*. University of Birmingham School, Birmingham, UK (2005)

[3] Bergen, B.: *Louder Than Words*. Basic Books, New York, NY (2011)

[4] Black, M.: *Models and Metaphors*. Cornell University Press, Ithaca, NY (1962)

[5] Desai, R., Binder, J., Conant, L., Mano, Q., Seidenberg, M.: The neural career of sensory-motor metaphors. *Journal of Cognitive Neuroscience* **23**, 2376–2386 (2011)

[6] Lakoff, G.: Metaphor and embodiment: How the neural theory of metaphor explains embodied cognition results. urs https://www.ualberta.ca/iclc2013/ABSTRACTS/Lakoff.pdf (2013). Accessed 18-6-15

[7] Lakoff, G., Johnson, M.: *Philosophy in the Flesh: The Embodied Mind and Its Challenge to Western Thought*. Basic Books, New York, NY (1999)

[8] MacKay, D.: *Information, Mechanism and Meaning*. MIT Press, Cambridge, MA (1969)

[9] McNerney, S.: A brief guide to embodied cognition: Why you are not your brain. url http://blogs.scientificamerican.com/guest-blog/a-brief-guide-to-embodied-cognition-why-you-are-not-your-brain (2011). Accessed 17-6-15

[10] Narayanan, S.: *Karma: Knowledge–based active representations for metaphor and aspect*. Ph.D. Dissertation (1997)

[11] Panikkar, R.: *The Rhythm of Being*. Orbis Books, Maryknoll, NY (2010). Based on the Gifford Lectures, 1989

[12] Paul, A.: Metaphor and the bounds of expression. *Philosophy and Rhetoric* **5**, 143–158 (1972)

[13] Searle, J.R.: Metaphor. In: A. Ortony (ed.) *Metaphor and Thought*. Cambridge University Press, London (1979) pp. 92–123

Index

FSC
www.fsc.org
MIX
Papier | Fördert
gute Waldnutzung
FSC® C083411

Kolibri 360 GmbH
Ferdinand-Jühlke-Straße 7
99095 Erfurt, Deutschland
produktsicherheit@kolibri360.de